Thomas Panhofer

Self-Healing Integrated Circuits

AF060121

Thomas Panhofer

Self-Healing Integrated Circuits

Feasibility of Asynchronous Implementations

Südwestdeutscher Verlag für Hochschulschriften

Impressum/Imprint (nur für Deutschland/only for Germany)
Bibliografische Information der Deutschen Nationalbibliothek: Die Deutsche Nationalbibliothek verzeichnet diese Publikation in der Deutschen Nationalbibliografie; detaillierte bibliografische Daten sind im Internet über http://dnb.d-nb.de abrufbar.
Alle in diesem Buch genannten Marken und Produktnamen unterliegen warenzeichen-, marken- oder patentrechtlichem Schutz bzw. sind Warenzeichen oder eingetragene Warenzeichen der jeweiligen Inhaber. Die Wiedergabe von Marken, Produktnamen, Gebrauchsnamen, Handelsnamen, Warenbezeichnungen u.s.w. in diesem Werk berechtigt auch ohne besondere Kennzeichnung nicht zu der Annahme, dass solche Namen im Sinne der Warenzeichen- und Markenschutzgesetzgebung als frei zu betrachten wären und daher von jedermann benutzt werden dürften.

Coverbild: www.ingimage.com

Verlag: Südwestdeutscher Verlag für Hochschulschriften GmbH & Co. KG
Heinrich-Böcking-Str. 6-8, 66121 Saarbrücken, Deutschland
Telefon +49 681 37 20 271-1, Telefax +49 681 37 20 271-0
Email: info@svh-verlag.de

Approved by: Vienna, Vienna University of Technology, Dissertation, 2012

Herstellung in Deutschland (siehe letzte Seite)
ISBN: 978-3-8381-3366-9

Imprint (only for USA, GB)
Bibliographic information published by the Deutsche Nationalbibliothek: The Deutsche Nationalbibliothek lists this publication in the Deutsche Nationalbibliografie; detailed bibliographic data are available in the Internet at http://dnb.d-nb.de.
Any brand names and product names mentioned in this book are subject to trademark, brand or patent protection and are trademarks or registered trademarks of their respective holders. The use of brand names, product names, common names, trade names, product descriptions etc. even without a particular marking in this works is in no way to be construed to mean that such names may be regarded as unrestricted in respect of trademark and brand protection legislation and could thus be used by anyone.

Cover image: www.ingimage.com

Publisher: Südwestdeutscher Verlag für Hochschulschriften GmbH & Co. KG
Heinrich-Böcking-Str. 6-8, 66121 Saarbrücken, Germany
Phone +49 681 37 20 271-1, Fax +49 681 37 20 271-0
Email: info@svh-verlag.de

Printed in the U.S.A.
Printed in the U.K. by (see last page)
ISBN: 978-3-8381-3366-9

Copyright © 2012 by the author and Südwestdeutscher Verlag für Hochschulschriften GmbH & Co. KG and licensors
All rights reserved. Saarbrücken 2012

Contents

1 Introduction — 1
   1.1 Motivation — 3
   1.2 Contribution and Objectives — 3
   1.3 Structure of the Thesis — 4

2 Principles of Fault Tolerance and Asynchronous Logic — 7
   2.1 Basics of Fault Tolerance — 7
      2.1.1 Terminology — 7
      2.1.2 Fault Classification — 9
      2.1.3 Fault Models — 9
      2.1.4 Masking Effects — 10
      2.1.5 Fault Hypothesis with Respect to this Thesis — 10
   2.2 Increasing Circuit Reliability — 11
      2.2.1 Introduction — 11
      2.2.2 Methods — 11
   2.3 Introduction to Asynchronous Logic — 14
      2.3.1 General — 14
      2.3.2 Classification of Asynchronous Circuits — 15
      2.3.3 Asynchronous Protocols — 15
   2.4 Four-State-Logic — 16
      2.4.1 General — 16
      2.4.2 Combinational Logic — 17
      2.4.3 Registers — 18
      2.4.4 Timing Parameters — 19
      2.4.5 Faults in Asynchronous Circuits — 21

3 State of the Art of Circuit Reconfiguration — 23
   3.1 Introduction to Autonomous Self-Repair — 23
   3.2 Circuit Reconfiguration — 25
      3.2.1 Runtime Reconfiguration with FPGAs — 25
      3.2.2 Dynamic Rotation and Free for Test — 26
      3.2.3 Fine-Grained Self-Healing Hardware — 28
      3.2.4 Dynamic Reconfiguration using Atomic Fault Tolerant Blocks — 29
      3.2.5 Column-Based Precompiled Configuration — 30
      3.2.6 Roving STARS — 31
      3.2.7 Node Covering Technique — 32

	3.2.8	Method of Shifting Configuration Data	33
	3.2.9	Nature-Inspired Methods	34
	3.2.10	Self-Repair using Re-Configurable Logic Blocks (RLBs)	35
	3.2.11	Self-Healing Asynchronous Arrays	37
3.3		Comparison and Conclusion	38

4 Self-Healing Approach — 41
- 4.1 Introduction 41
- 4.2 Architecture Overview 42
  - 4.2.1 Concept and Fault Hypothesis 42
  - 4.2.2 Fault Locations 43
  - 4.2.3 Reconfiguration of Combinational Logic 44
  - 4.2.4 Reconfiguration of Control Logic 45
  - 4.2.5 Reconfiguration Unit 46
  - 4.2.6 Self-Healing Reconfiguration Unit 48
  - 4.2.7 Default State of Configurable Elements 49
  - 4.2.8 Transformation from FSL to SH-FSL 50
- 4.3 The Principle of Pipeline Reconfiguration 51
  - 4.3.1 Introduction 51
  - 4.3.2 Faults at SHC Inputs 51
  - 4.3.3 Faults at Register Inputs 53
  - 4.3.4 Faults at Acknowledge Signals 53
  - 4.3.5 Timing Investigation of Different Pipeline Configurations 53
  - 4.3.6 Summary 57
- 4.4 Fault Diagnosis in a Pipeline 58
  - 4.4.1 General 58
  - 4.4.2 Fault at SHC Input 59
  - 4.4.3 Fault at Register Input 59
  - 4.4.4 Fault at Acknowledge Signal (Pass Input) 60
  - 4.4.5 Observable Symptoms 60
  - 4.4.6 Effects due to Multiple Faults 65
  - 4.4.7 Summary 66
- 4.5 Reconfiguration Rules and Algorithm Efficiency 67
  - 4.5.1 Reconfiguration on Pipeline-Level 67
  - 4.5.2 Embedded Fine-Granular SHC Reconfiguration 68
- 4.6 Overhead of the Self-Healing Approach 71
- 4.7 Annex 74

5 Analysis, Simulations and Experimental Results — 77
- 5.1 Introduction 77
- 5.2 Environment 78
  - 5.2.1 General 78
  - 5.2.2 Simulation Environment for Pipeline Reconfiguration 78
  - 5.2.3 Environment for Hardware Experiments 81
- 5.3 Reconfiguration of Self-Healing Cells 83
  - 5.3.1 Fault Tolerance of Fine and Coarse Granular Self-Healing Cells 83

Contents

- 5.3.2 Optimization of Self-Healing Cells 84
- 5.4 Simulation of Pipeline Reconfiguration 89
  - 5.4.1 Simulation of a Deadlock Recovery 89
  - 5.4.2 Simulation Results of Different Reconfiguration Algorithms 91
  - 5.4.3 Result Summary and Comparison 100
- 5.5 Hardware Fault Injection Experiments 103
  - 5.5.1 General 103
  - 5.5.2 Results of Pipeline without Acknowledge Switches 105
  - 5.5.3 Results of Pipeline with Acknowledge Switches 108
  - 5.5.4 Summary 109
- 5.6 Hardware Implementation of a Complex Self-Healing Circuit 110
  - 5.6.1 Introduction 110
  - 5.6.2 The GAIA Pre-Processing Algorithm 110
  - 5.6.3 FSL Implementation 112
  - 5.6.4 Self-Healing GAIA Algorithm 114
  - 5.6.5 Results 116
- 5.7 Reliability Analysis of Self-Healing Circuit 118
  - 5.7.1 Introduction 118
  - 5.7.2 GAIA Reliability Estimation 119
- 5.8 Comparison with State-of-the-Art Methods 121
  - 5.8.1 Duplex System 121
  - 5.8.2 TMR System 121
  - 5.8.3 Comparison with Self-Healing Approach 121
- 5.9 Summary and Discussion 123
  - 5.9.1 Fault Frequency 124
  - 5.9.2 Timing Assumptions 124
  - 5.9.3 Mis-alignment of Nominal and Redundant Path 124
- 5.10 Annex 126

6 Conclusion and Outlook 131

Chapter 1

Introduction

The complexity of applications is continuously increasing and at the same time the requirements on power consumption, reliability and performance become more and more stringent. The semiconductor industry is following the demands from application side very well and offers more resources, higher operating speed and less power consumption at lower supply voltage each new chip generation. In 1965 Gordon Moore [91] predicted a duplication of resources per square inch every 18 months[1]. More than 40 years later the International Technology Roadmap for Semiconductors (ITRS) [58] shows that this prediction is still fairly true and, as can be seen in Figure 1.1 [119], has not yet reached the end.

Although this evolution was and is one of the major contributors for the level of technology we have reached, the designers are also facing new challenges:

1. Susceptibility to Faults: Due to the lower charge stored at the circuit nodes the chips become more prone to various fault effects. For example, in the past transient faults like Single-Event-Upsets (SEUs) were a serious problem only in extreme environments such as space, but now become problems even in on-ground applications [63, 94, 46, 128]. Apart from soft-errors, the increased technology scaling causes more and more manufacturing imperfections [93, 110, 39]. On the one hand, this decreases the yield and thus increases the component costs. On the other hand, some of these faults might even not be detected during manufacturing tests, and may appear as permanent faults even after a long time of successful operation. So, recently also permanent defects gained higher attention [4, 72]. Moreover, the single-fault model seems to become outdated. The probability for multiple errors caused by a single fault event is increasing, and a massive number of defects has to be assumed [1].

2. System Complexity: The increasing complexity of the applications makes system failures more difficult to predict and to understand. Thus, they are often not considered sufficiently in the system design. Furthermore, new and powerful technologies and semiconductor processes are often either not available or not tolerated, respectively, in conservative industries (such as space), or are not robust enough - and the available robust technologies typically lack in performance. For example, radiation-tolerant anti-fuse FPGAs offer less than 1/10 of the resources compared to SRAM-based FPGAs [40]. As a consequence, the components selected for such applications often have to be operated at their limits.

---
[1]The time horizon varies between 12 and 24 months but seems to settle at 18 months.

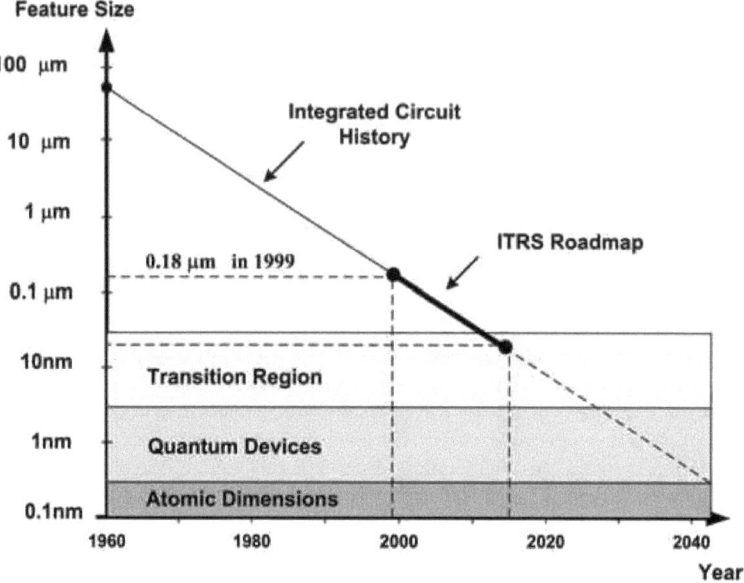

Figure 1.1: Trend of Feature Size [119]

3. Electromagnetic Field (EMF) Pollution: The operation environment becomes harsher due to increasing EMC disturbances caused by the huge number of wireless services (WiFi, Bluetooth, mobile communication, GPS, etc.).

4. Extreme Application Demands: The progress in technology causes new areas of applications to evolve. For example, interplanetary space missions [22] where the spacecraft is "on the way" for several years, require a highly robust circuit and autonomous failure detection, isolation and repair (FDIR), as a manual repair is not possible.

5. Use of complex integrated circuits in critical applications: While failing systems and subsystems due to defective components might be tolerated to some extent in consumer electronics, this is absolutely not acceptable in applications where human life is jeopardized or where human interaction (e.g. repair) is not possible or extremely expensive. Typical examples for the latter ones are automotive, aerospace and space applications.

This list is certainly not complete, but illustrates some important issues that need particular attention.

In conservative industries, e.g. the space industry, the traditional approach to gain high reliability is *fault avoidance* by using screened components and reducing component stress by parameter derating. Of course, high quality components are very expensive, and nevertheless there is a limit that can only be overcome by additional means. Rather than trying to make the components perfectly reliable, the trend now changes to build reliable systems from unreliable components [4] – which requires *fault tolerance* techniques.

1.1 Motivation

The traditional approach to increase fault tolerance of unreliable modules is to apply hardware redundancy on system level [115]. The granularity is typically rather coarse and thus quite expensive in terms of hardware costs. In addition, dedicated fault control mechanisms are required, which regularly check some characteristics of the system and decide, when the redundant section shall be used. A significant amount of resources usually has to be spent on the management of the implemented fault tolerance mechanisms.

Each new generation of programmable logic devices, such as FPGAs and ASICs, provides more and more resources, so that these components are increasingly used to accomplish the main functionality of an application. However, as they comprise the functionality of a huge system of earlier days in a single chip, this implies that they become critical elements in the system. Although their failure rate is rather low [16], in particular for long mission times the probability for even multiple faults can become an issue. While a lot of methods for handling transient faults are well established, there are hardly any concepts available to deal with permanent faults.

From an application point of view there is a demand on a generic and reliable hardware platform, which can handle transient as well as permanent faults, and which is able to autonomously recover from faults without external interaction. Runtime reconfiguration is deemed to be a promising way to provide such features.

Methods applied on transistor level are considered to be too specific, and furthermore would require completely new component libraries. The solution to be found should therefore apply on gate level, which will also ease the development and prototyping.

Irrespectively of the implementation, some kind of diagnosis is required first to locate the fault. Here asynchronous logic has significant advantages, as it is inherently tolerant against timing variations e.g. caused by changed circuit routings, and it tends to stop operation in the presence of permanent faults [3]. Although asynchronous logic is not well established at least for large applications and lacks from appropriate design tools, it was already shown that it is possible to implement reasonably complex asynchronous circuits such as processors even in standard FPGAs [14].

The main drawback of asynchronous circuits is the inherent logic overhead due to the design technique, e.g. handshake protocols, dual rail encoding, etc. However, by improving the design flow it is possible to reduce the overhead significantly [26]. Anyway, the hardware overhead is not seen as major criterion for this thesis.

1.2 Contribution and Objectives

This thesis describes an adaptive system, which is able to autonomously perform all necessary tasks to stay alive, while the higher level application can rely on a correctly working hardware and can concentrate on the intended function instead of dealing with fault tolerance.

The concept implements *autonomous self-repair within integrated circuits on gate level* and can be based on existing processes and design tools. This requires to decompose the circuit into parts with appropriate granularity, which are then implemented by flexible, reliable structures. These elements allow being reconfigured in case errors occur to bypass the defective resources with working ones. The concept is based on asynchronous logic and makes use of the inherent properties, such as fail-stop behavior. The reconfiguration is performed during runtime and is transparent to the application. The only impact is increased delay, which is tolerated by definition of an asynchronous circuit.

The methodology presented in this thesis is called *self-healing*, as the aim is to achieve a completely autonomous handling and repair of faults occurring during operation. The main objectives can be summarized by the following items:

- Recover from multiple permanent faults and errors
- occurring in integrated circuits, irrespectively of their origin
- within a predictable timing
- with deterministic measures
- by autonomous reconfiguration
- transparent to the application
- using existing processes and standard libraries

Although the considerations in this thesis are not focused on a particular application, harsh environment and high reliability as e.g. needed in space applications, are always kept in mind regarding their special requirements. This implies e.g. that the circuit behavior shall be deterministic and reproducible so that it can be well analyzed although the circuit architecture might evolve during operation.

To achieve these goals a novel structure of self-repairing blocks was developed. Three important topics, namely fault detection, diagnosis and repair, are of particular interest and needed to be harmonized. The behavior of asynchronous pipelines in the presence of permanent (multiple) faults is investigated in detail and various reconfiguration options are analyzed. An architecture as well as algorithms for fault removal are presented and the optimum reconfiguration strategy is elaborated. The effectiveness and suitability of the concept is proven by simulations and hardware experiments.

Within the scope of this thesis several generic tools as well as simulation and analysis models were established, which allow comparing different implementations, architectures and algorithms in a structured and deterministic way.

1.3 Structure of the Thesis

Chapter 2 briefly describes the terminology used in this thesis and gives an overview about the considerations for fault tolerant systems on chip level. Furthermore, the basic principle of asynchronous logic in general and the used design style *Four State Logic (FSL)* in particular is described.

Chapter 3 presents the state-of-the-art of circuit reconfiguration in integrated circuits. The concepts are presented and evaluated with respect to the requirements and needs defined for this thesis.

1.3. Structure of the Thesis

Chapter 4 presents details about the developed architecture and the methodology for circuit reconfiguration. It is explained how to apply the concept to a standard FSL circuit. The behavior of FSL pipelines in failure cases and the respective observable symptoms are derived, which builds the prerequisite for establishing an appropriate reconfiguration algorithm later on. Lastly, the overhead of the self-healing approach is assessed.

In **chapter 5** the results of all simulations, analysis and experiments as well as a description of the used tools, development and prototyping environments is presented. Different reconfiguration algorithms are implemented in a simulation model as well as in hardware and compared with each other. The proof-of-concept is established by - among several other simulations and experiments - the implementation of a complex video processing circuit and by hardware fault injection experiments. A reliability analysis is performed for an exemplary circuit to evaluate the gain of reliability for the developed concept. The results are compared with other state-of-the-art methods.

Finally, **chapter 6** forms the conclusion. The work is compared with the initial goals and an outlook to future work is given.

Chapter 2

Principles of Fault Tolerance and Asynchronous Logic

2.1 Basics of Fault Tolerance

This section presents an overview and clarification of commonly used terms related to reliability and fault tolerance. Details can be found e.g. in [2].

2.1.1 Terminology

Errors may appear unpredictably in every electronic circuit. The method, how to cope with the error and to which extent this has to be done, depends on the application and its criticality in the system. The threats to a system can be distinguished into three main categories: *faults*, *errors* and *failures* [73].

- The term *fault* is used to describe the *cause of an error* in a system. A fault is an action or event that affects a system in a way that an error can occur, however, the existence of faults does not necessarily imply that an error occurs. Possible faults could be e.g. single event effects due to radiation.

- *Error* describes an unintended system state due to a manifested fault. Errors may be explicitly related to external events (e.g. EMC or radiation) or e.g. due to defective components, aging effects of components inside the chip, etc. The existence of an error still does not necessarily imply that the system fails. If the error is e.g. located in unused resources or in a currently unused circuit part, it will not affect the circuit functionality, at least until this part or these resources come into operation. An error that has not yet triggered a failure is called *latent*.

- *Failure* is the effect visible from the outside world, i.e. a deviation from the system specification caused by an error that has propagated to the system boundary. This could be e.g. a wrong result or a result produced at a wrong (unexpected) time. Failures can already occur at component level, e.g. if a transistor does not fulfil its specified performance any more. A failure in a subsystem might become a fault in the higher level system.

The causal relationship between these terms and the propagation between system boundaries can be expressed by the "fundamental chain" [2]:

$$... \rightarrow failure \rightarrow fault \rightarrow error \rightarrow failure \rightarrow fault \rightarrow ...$$

A system should be designed so that no *single-point failure* exists, i.e. that no single fault/error causes the system to deviate from the specification. Typically, this is analyzed in a *Failure Mode, Effects and Criticality Analysis* (FMECA) [95].

While some effects can be handled by preventive actions aiming for *fault avoidance* (e.g. shielding to reduce radiation dose), others require appropriate *fault and error detection* techniques to be able to detect and repair the error before it propagates in the system and results in a failure, i.e. *fault tolerance*. The method to be applied depends on the application and on its criticality in the system.

Doumar et.al. [18] further introduces the term *defect tolerance*, which is used to describe techniques performed by the manufacturer, i.e. handles defects that occur during manufacturing, whereas fault tolerance describes the method done by the user to heal errors which occur during operation of the circuit.

Before a fault can be corrected, it must be detected and correctly identified. There exist several ways of error detection, a very common example in particular for serial transmissions is parity information [115].

The aim of fault tolerance is to provide the requested service even in the presence of faults, i.e. the system must be able to re-construct the information transmitted by the sender from the received faulty data. Of course, fault tolerance mechanisms can only correct faults and errors that have been considered in the fault hypothesis. Typically, the limitation is the number of faults and the fault frequency. For example, memories can be protected by an *Error Detection and Correction* (EDAC) mechanism. Depending on the number of syndrome bits (which limit the useable memory size) one or more faults can be detected and corrected [115]. With the EDAC an error may be removed before it affects the system.

The term *dependability* summarizes the attributes *reliability, availability, maintainability, safety, integrity* and *confidentiality* [2]. Except of reliability and availability the attributes are rather qualitative and cannot be quantified by measurements. Reliability describes the probability that a system provides its specified service at a particular time. Availability gives the relation between the time the system fulfils its specification and the time the system is repaired (maintenance time), i.e. the readiness of usage. For systems without repair (e.g. typically space systems) the reliability defines the mission success, as the system is considered to be failed after the first fault occurrence. An important parameter for reliability analysis is the failure rate of the used elements, which expresses the probability of failure per time. In fact, components cannot have a failure rate, as they cannot be repaired, but anyhow the term failure rate is commonly used for components to express the contribution to the system failure rate.

The time until a system fails the first time is called *Mean Time To Fail* (MTTF). For repairable systems a *Mean Time Between Failure* (MTBF) can be defined (MTBF = 1/failure rate), as well as a *Mean Time To Repair* (MTTR), i.e. the average time it takes to repair a system. Obviously, in order to be become practically usable, the MTTR must be significantly lower than the MTBF.

2.1. Basics of Fault Tolerance

2.1.2 Fault Classification

Faults can be classified into different categories [2], of which the distinction by their persistence into *temporary* and *permanent* faults is of particular interest for this thesis.

Transient faults are temporary faults originating from the physical environment. They are present for a limited time and can be caused e.g. by radiation [24] or by *Electromagnetic Interference* (EMI). Radiation induced faults are also called *Single Event Effect* and can be further classified into *Single Event Transient* (SET), *Single Event Upset* (SEU), *Single Event Latchup* (SEL) and *Single Event Burnout* (SEB).

In general, radiation induced particle strikes change the electrical charge stored at a circuit's node and thus the voltage of this node. Depending on the amount of charge stored and removed/induced, as well as the circuit's technology and the driver's strength, the logic level of the signal is changed. A SET is a logic transition that is restored by the driver. If the changed logic level is stored and thus remains, the effect is called SEU. A SEU can be removed by re-defining the signal state or resetting the storage element. The duration of radiation induced transient faults is in the order of 1ns [17]. SEUs do not alter the hardware, and are thus defined as *soft error* [116].

Permanent faults typically model physical defects. Possible causes are manufacturing imperfections, overstress, electromigration [25], or can be a consequence of transient faults. Sometimes a particle strike triggers a parasitic thyristor and the signal remains in the wrong state, unless the device is power-cycled. This effect is called SEL. It is potentially destructive because high currents can be induced, leading to a SEB, which may destroy the device. The SEB was first seen in power MOSFETs and causes a physical (hardware) defect of the device due to high current. It is thus defined as *hard error*. As hardware defects cannot be removed, they lead to *permanent errors*. The effect in MOSFETs is also called single event gate rupture (SEGR). A similar effect has been observed in CMOS circuits and is there called single event dielectric rupture (SEDR) [38].

In SRAM based FPGAs these definitions become a bit indistinct. As the function is defined by a bitstream stored in a SRAM, a soft error can change the circuit structure, which would be visible as permanent error, although the hardware is not defect (i.e. a permanent soft error). Such an error could be resolved by restoring the original function e.g. by scrubbing or a device reset.

2.1.3 Fault Models

A very popular model to describe the fault behavior of a circuit under test is the *(single) stuck-at fault model* (SSAF) [64, 103, 39]. According to this fault model a circuit line is stuck-at one or zero if it is disconnected from any other circuit's wires and connected to the power supply or ground.

Although it is a very simple model, it covers at least 70% of fabrication defects [76] and is well suited to model permanent faults. In contrast to "open" faults, i.e. simply disconnecting a signal from its target, stuck-at faults represent shorts and thus need particular handling with respect to fault isolation.

Bridging faults model connections to other signals, i.e. the logic state is controlled by another signal that is connected due to a physical fault [39].

Since some fault effects could be masked by the simple stuck-at fault model or the bridging model (if the fault forces the signal to the same logical value as it has anyhow), some other fault models have been developed.

The *bit-flip* model simply inverts a signal state. This can be troublesome as e.g. for latent faults the signal could toggle its state several times, which does not represent the correct physical behavior. Other fault models such as *delay faults* or *pulse faults* do not affect the circuit's logical function and are only relevant for the investigation of transient fault behavior, which is out of scope for this thesis.

2.1.4 Masking Effects

Faults might be masked and thus prevented from becoming active due to mainly the following three reasons [114]:

- Temporal masking: The fault does not affect the circuit function because it appears at a time where the signal is not evaluated (e.g. between clock edges).

- Electrical masking: The fault is attenuated sufficiently by gates, wires, etc. so that it does not change a signal state.

- Logical Masking: The logical function receiving a faulty input is insensitive to the signal state, e.g. an OR gate does not change its output if one input has state 1 and the fault forces the other input to 0.

The only relevant masking effect in the scope of this thesis is the logical masking, which could delay the occurrence of inconsistent data.

2.1.5 Fault Hypothesis with Respect to this Thesis

The fault hypothesis summarizes all conditions, under which faults are assumed to occur, and forms the basis for fault handling in an application.

Within this thesis the following fault hypothesis applies:

- The faults are modeled according to the stuck-at fault model. Both stuck-at-1 and stuck-at-0 faults may occur.

- Only permanent faults are considered, irrespectively of their origin. Whenever the term *fault* is used, it refers to a *permanent fault*.

- Multiple faults are considered with the restriction that only one fault occurs at a time. As permanent faults are considered, they are all existent simultaneously in the system at the end, but there was sufficient time to repair a fault before the next one occurred.

- Faults are modeled at register level. With this restriction the internal design of basic elements need not be taken into account, and thus the faults can be simulated at different platforms and target devices (e.g. FPGA design vs. ASIC design).

2.2 Increasing Circuit Reliability

2.2.1 Introduction

Ideally, a circuit shall never fail after it is switched on, i.e. have a reliability of 100%. Clearly this is not possible, as components degrade over time and the environment influences various parameters that might reduce the lifetime. In order to increase the reliability of a circuit, either faults have to be avoided (fault avoidance) or the application must be able to deal with faults (fault tolerance).

There exist basically three methods to improve the reliability of a circuit/system.

1. Reduce the stress the components are exposed to in order to reduce the failure rate and to add additional design margin (fault avoidance)

2. Use high quality (screened) components offering higher reliability and robustness against faults (fault avoidance)

3. Implement redundancy in the circuit (fault tolerance)

In the following sections these methods are described in more detail.

2.2.2 Methods

Component Stress

The lifetime, reliability and performance of components are affected by several factors, such as e.g. voltage/current stress and temperature. Temperature is one of the main contributors that decrease the component reliability [16], and the relation between part failure rate and device temperature can be expressed by the Arrhenius law [62].

A very common approach to reduce the stress is to *derate* the design parameters, having the advantage that the higher reliability is "built-in" and does not need any further actions during operation. Derating rules and methods for analysis have been established and standardized by the industry, such as by the "European Cooperation for Space Standardization" [27], so that it's basically straightforward to provide evidence for the compliance to these rules. The drawback of component derating is that oversized components might be needed, which require more board space, are more expensive and often also have higher power consumption.

Recently methods for thermal de-stressing in integrated components [70] have been published, where the circuit is reconfigured to use redundant resources to reduce the stress of e.g. clock trees. Currently no results of field experiments are available to evaluate the practicability and gain in reliability in real applications.

Component Quality

Component screening means to expose components to stringent tests to find those with outstanding performance even under extreme conditions. It is obvious that such components are very expensive and have long lead times.

Particular processes have to be used to harden components against radiation effects. Unfortunately, these components often provide reduced electrical performance and have higher power consumption. Since dedicated design measures have to be applied, modern components

2. Principles of Fault Tolerance and Asynchronous Logic

are rarely available in hardened technology. Furthermore, such components are often subjected to export regulations which increases both lead time and cost. Basically, the trend currently goes towards building highly reliable systems from less reliable but cheaper components [4].

As for the parameter derating, the higher reliability is "built-in" and does not need to be handled during operation. Software tools for reliability calculations according to established models, such as MIL-HDBK-217 [16], allow to assign the component quality for standardized screening so that the effect on the system can be analyzed very easily.

Redundancy

Redundancy means to add additional information to a data path so that it is possible to determine at least the correctness of data, or even to correct it in case of errors. Basically, redundancy can be applied (i) in time (e.g. dual calculation), (ii) in the value domain (e.g. correcting codes) or (iii) in hardware. While time redundancy and codes are well suited to correct effects caused by transient faults, only hardware redundancy is able to handle permanent errors. Subsequently, whenever the term "redundancy" is used, it refers to hardware redundancy.

In hardware redundancy alternate paths in a circuit/system are implemented that take over in case of a failure, or provide additional information that is used by voters to take a decision on the correctness of results.

Redundancy can be distinguished into *active redundancy* (Figure 2.1a) and *standby redundancy* (Figure 2.1b). In active redundant systems the spare resources are continuously powered simultaneously with the nominal circuit, e.g. two power supplies that are ORed with diodes. A particular active redundant system is *N-modular redundancy* (Figure 2.1c), where the majority m-out-of-n redundant results is used. The most common and widely accepted majority voting system is the *Triple Modular Redundancy* (TMR) structure, where the results of three paths are evaluated by a majority voter.

In a standby redundant system the spare elements are by default powered down and put into operation when the nominal circuit fails. Standby redundant systems need a dedicated fault detection logic and a controller which performs all necessary boot actions (e.g. transferring internal states), i.e. the circuit/system will be down for some time. A redundant system with one parallel path is also called *duplex system*, and is usually used in standby redundancy.

Redundancy can be implemented on different levels and in different granularity. In general, component redundancy is superior to system redundancy ([115], chapter 3.3) because the finer granularity implies more alternate paths[1]. However, implementing redundancy is a complex task and not all theoretical solutions can be realized. Switches, voters and redundancy controllers significantly contribute to the system reliability, as they add series elements in the reliability path and thus lower the overall reliability. Particular techniques might be necessary, e.g. to split the triplicated logic and the voter into different components [121]. Furthermore, components can not always be parallelized (e.g. two brake pedals in a car would be an impractical solution; also electrical components cannot simply be put in parallel). The impact of the series elements and additional interconnect resources also puts practical limits on the granularity of redundant systems [1].

---

[1] Note that "component" does not necessarily mean "electrical component" in this context. A component could e.g. be a turn indicator in a car.

2.2. Increasing Circuit Reliability

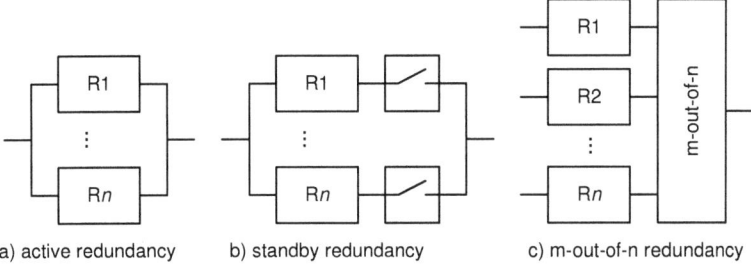

Figure 2.1: Redundancy Techniques

Finally, it must be considered that detection, diagnosis and repair circuits add additional resources to the application, that itself are susceptible to faults and thus limit the achievable reliability.

Triple Modular Redundancy (TMR) is one of the most popular techniques and can be implemented on gate as well as an system level [109]. Tools are available that assist the implementation during the design process [12, 13]. Such structures can even be implemented e.g. in flip-flop resources in FPGAs [8] to make them tolerant against SEUs.

Typically, TMR is well suited to increase the tolerance against transient faults, as the result will be correct without delay and additional diagnosis. As in case of SRAM FPGAs a SEU can change the circuit structure, a possible approach is to triplicate the design (three identical FPGAs) and use an external voter [109]. If an error is detected, the correct FPGA configuration can be recovered by partial reconfiguration [131, 132] without affecting the functionality of the operating circuit. Similar approaches are described in [122] and [121]. In [121] three soft CPU's are implemented in an SRAM based FPGA and are voted externally by a rad-had antifuse FPGA. In case of an error the SRAM-based FPGA is scrubbed to recover from the upset. However, the internal states are lost by this action. The Maxwell Super Computer for Space SCS750 [122] contains three non-rad-hard PowerPC's and a rad-hard voter FPGA to detect the errors and control the recovery actions. Once an error occurs, the processor registers are stored, the processors are reset and then the old register contents are restored to be able to start from the previous state. In [67, 77] TMR based approaches combined with time-redundancy are presented, in order to reduce the area overhead. The results show that it depends both on the type of data structure that shall be protected (throughput logic, state-machine logic, I/O logic, special features) and the location of the voter (only at the last output, inside the design) how efficient and costly the implementation is.

For permanent faults TMR soon reaches its limits as after two faults in different redundancy elements it is not possible any more to correctly conclude on a majority. TMR in its basic variant is thus only effective for short mission times unless repair is possible [115, 134]. Various methods for TMR systems to improve the long-time reliability are proposed in [36, 35, 29, 30], however, they can only handle soft errors in the configuration memory but not real hardware defects, so these approaches are not considered any further within this thesis. To lengthen the mission time, it is necessary to handle permanent errors by changing the circuit structure and replacing defective elements by working ones, e.g. as proposed in [65]. In [129] an evolutionary voting system is described, where a genetic algorithm tries to bring a faulty module back to operation. The results indicate that this approach could be useful, however, only simulations with a simple model of an FPGA have been performed.

In synchronous circuits implementing a TMR system is straight-forward, as all results are available at the same time instant. TMR in asynchronous circuits is a bit more complex, as a fault in the asynchronous logic could prevent the result to appear at the output and thus block the comparison permanently [102]. In [41] a modified TMR architecture is proposed that can be used also for asynchronous circuits.

2.3 Introduction to Asynchronous Logic

This chapter gives an overview about asynchronous logic in order to understand the described self-healing concept later on. For details the reader is referred to e.g. [47, 92, 120, 78].

2.3.1 General

Digital circuits can be distinguished into two significant areas - *synchronous logic* and *asynchronous logic*. While any transition in synchronous logic is triggered by a clock, which thus defines the timing of the whole circuit, asynchronous logic relies on other mechanisms.

One big advantage of the discrete timing in synchronous logic is that the logic states are only relevant at a clock edge. Between the clock edges the signals may enter other states, e.g. due to faults, without having any effect on the circuit functionality. This property simplifies the circuit design, but implies a restriction on the maximum reachable speed. The circuit must be designed to achieve the design goal even under worst case propagation delays, setup and hold times.

Asynchronous circuits do not require a clock, instead a local handshake provides the information that new data is available and ready to be processed. The speed of the circuit is therefore determined by the propagation delay of the involved elements, and data is processed when it is available, and not at discrete times.

Apart from several other benefits compared to synchronous logic, such as lower power consumption [127, 81] and lower electromagnetic emissions [81, 99], the main advantage of asynchronous logic with respect to this thesis is the inherent robustness against faults. Due to the insensitivity to circuit timings, a lot of environmental effects, such as thermal drifts, do not influence the behavior of the design.

Although this basically sounds excellent for electronic circuits, asynchronous logic also has some drawbacks. Currently, there exist no comfortable design tools, the design itself is more complex and it requires more resources than synchronous logic. Furthermore, although research on asynchronous FPGAs is ongoing [123, 48], there are no commercial prototyping environments available, which makes integration tasks very difficult. As will be shown later, it is possible to use standard FPGAs to implement asynchronous designs, however, the results with respect to resource effort, timing and efficiency are not very representative, as these FPGAs are optimized for synchronous designs.

2.3. Introduction to Asynchronous Logic

2.3.2 Classification of Asynchronous Circuits

Asynchronous logic can be distinguished into two main models:

- the *bounded delay model*, which constrains gate and wire delays
- the *unbounded delay model*, which admits arbitrary delays at least for some parts of the circuit

Synchronous circuits follow the bounded delay model, as it is assumed that all transient states have settled to a steady state before the next clock edge occurs. Asynchronous circuits following the bounded delay model require particular timing assumptions and are also referred to as *self-timed* circuits [47].

Two families of asynchronous circuits following the unbounded delay model can be distinguished, which are the *speed independent* (SI) and *delay insensitive* (DI) circuits. While speed independent circuits assume positive but unknown delays in gates and zero delays on wires, delay insensitive circuits do not apply any restriction on delays in gates and wires. The family of delay insensitive circuits is, however, restricted to circuits consisting only of Muller-C gates and inverters [79], which limits its usability for practical applications. The class of *Quasi Delay-Insensitive* (QDI) circuits is a bit less restrictive, and allows unbounded delays with the exception of *isochronic forks*. These forks anticipate that all transitions starting at the root of a fork reach the end at the same time, i.e. the difference of the branch delay is negligible. This assumption is feasible from a practical point of view, and the class of circuits becomes much bigger. If all forks in a circuit are isochronic, the circuit can be considered to be speed independent. In a practical implementation, isochronic forks can be achieved on gate level (gates, registers), where matched delays are easier to control. The connection between such blocks is then delay insensitive [118].

2.3.3 Asynchronous Protocols

Without a clock, the validity and capturing of new data must be determined by some kind of *handshake protocol*. As shown in Figure 2.2, a *request event* is needed to inform the receiver that new data is available, and an *acknowledge event* to inform the sender that the data has been captured. During fault-free conditions these two events will alternate.

In synchronous circuits no handshake is needed, as the clock signal serves as global event that triggers any storage of data. In asynchronous logic different handshake protocols are possible, which can be distinguished e.g. by their encoding (level or transition) or by the number of protocol phases [53, 92, 117, 125, 86, 59].

The following section provides an overview about *Four State Logic* (FSL), also known as *Level Encoded Dual-Rail Signalling* (LEDR), which is the design style the self-healing concept is based on.

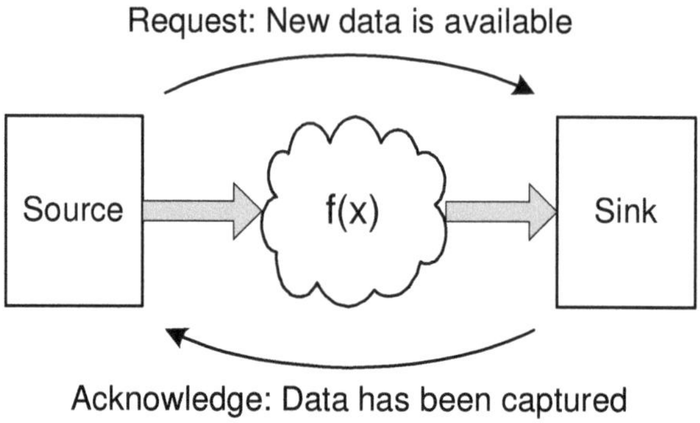

Request: New data is available

Source → f(x) → Sink

Acknowledge: Data has been captured

Figure 2.2: Handshake Principle in Asynchronous Circuit

2.4 Four-State-Logic

2.4.1 General

FSL is a *Quasi Delay Insensitive* asynchronous design style [118] that uses a two-phase handshake protocol. Consecutive data is separated by two alternating, diverse code sets $\varphi 0$ and $\varphi 1$, called *phases*. Figure 2.3 shows the encoding and transition between the boolean values TRUE/ FALSE denoted as 'h'/'l' in phase $\varphi 0$ and 'H'/'L' in $\varphi 1$. Each logic value is encoded by the two signal rails a and b [15]. A data vector is called *token*.

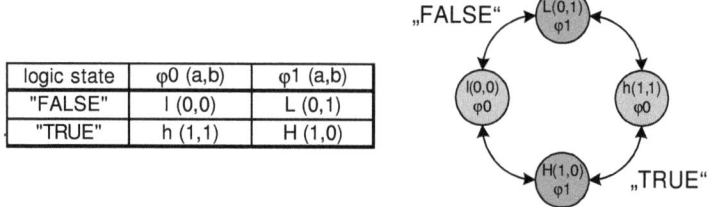

logic state	φ0 (a,b)	φ1 (a,b)
"FALSE"	l (0,0)	L (0,1)
"TRUE"	h (1,1)	H (1,0)

Figure 2.3: FSL Encoding and State Transitions

The data path of FSL circuits is modeled similarly as Sutherland's micropipeline [120] and is shown in Figure 2.4.

It comprises *capture-done registers* with combinational functions between the stages. The registers handle the handshake protocol, as they only capture and pass through new data if the subsequent stage has stored the current data.

The combinational functions between the particular pipeline stages are calculated by FSL gates, having an inherent synchronization mechanism: A new output is only generated when all inputs are in the same phase, otherwise the old output is preserved. This property is called

2.4. Four-State-Logic

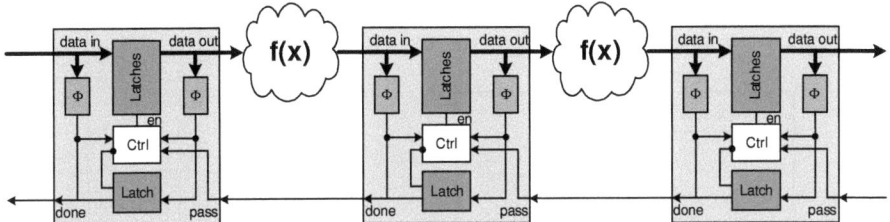

Figure 2.4: FSL Pipeline Structure

strongly indicating[2]. Thus, even a pure combinational function requires storage elements, which generates considerable area overhead compared to common single-rail logic.

An FSL based asynchronous embedded processor has already been successfully implemented [15].

2.4.2 Combinational Logic

FSL combinational functions can be designed with basic gates, such as AND, OR, XOR, etc. functions. With these basic gates larger and more complex circuits can be built by connecting them together such as with "normal" synchronous logic gates. As an example, an FSL AND gate is described in detail. Table 2.1 shows the truth table on signal levels for a 2-input FSL AND gate.

Z		E1			
		h	l	H	L
E2	h	h	l	hold	hold
	l	l	l	hold	hold
	H	hold	hold	H	L
	L	hold	hold	L	L

Table 2.1: 2-Input FSL-AND Truth Table (Signal Level)

If the inputs are consistent and have the same phase, the AND function is applied, and the output is generated in the same phase as the inputs. For all other cases the last valid output is preserved (designated with "hold"). This means that storage elements are needed in combinational FSL functions, as shown in Figure 2.5.

On rail level the truth table needs to be defined for two rails per signal (input). For each rail of the output a logic function is required that generates the respective set and reset signals for the RS flip-flops.

---

[2]In *weakly indicating* asynchronous logic single bits could already change, while the whole vector is still not valid.

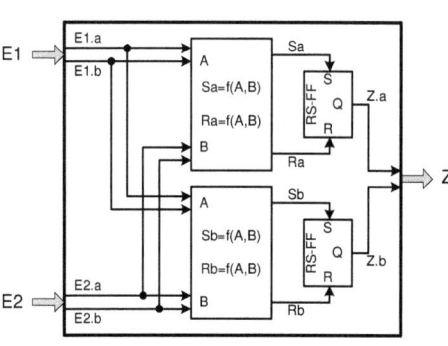

Figure 2.5: Schematic of 2-Input FSL AND Gate and Truth Table on Rail Level

2.4.3 Registers

An FSL register (Figure 2.6) contains phase detectors at the input and output and control logic for the handshake control. The phase detectors check that all input signals are within the same phase. If this is not the case, the phase detector output corresponds to the last phase. The input phase detector Φ enables the internal latches when all inputs are in the same phase, while the output phase detector freezes the latches after the complete data has been captured. Finally, an acknowledge signal informs the preceding stage to issue new data.

To store new data, (i) the phase of the latched data inside the register must differ from the phase of the data applied to the register inputs and (ii) the subsequent stage must have acknowledged that it is ready for new data, i.e. it has captured the last token.

In order to distinguish the handshake signals, the following terms are used within this thesis:

- *Done* is an output of the register and indicates that the register has captured new data. The logic state equals the phase of the captured data. This signal is fed towards the *Pass* input of the preceding register.

- *Pass* is an input to the register and indicates that the subsequent register has captured the last issued data. This signal is coming from the *Done* output of the subsequent register.

2.4. Four-State-Logic

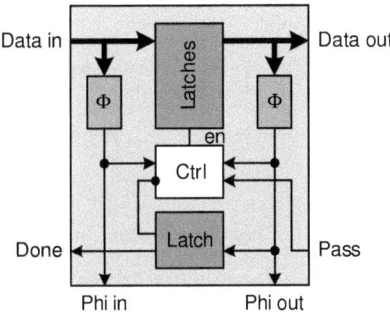

Figure 2.6: FSL Register

2.4.4 Timing Parameters

The following three timing parameters have to be considered in an FSL pipeline. Figure 2.7 shows the meaning of these parameters in a timing diagram.

- t_R is the time data needs to propagate from a register input to the output.
- t_H is the time between availability of new data at the register output until the *Done* signal is set accordingly.
- t_L is the time the data needs to propagate through the FSL logic $f(x)$.

A register will become ready to capture a new token if (i) it has captured the last valid token and (ii) it has received the corresponding acknowledge from the subsequent pipeline stage. The capturing process can be triggered by two conditions, depending on whether a new token or the acknowledge signal is provided earlier.

1. If the last captured token i was acknowledged to register n by the subsequent register $(n+1)$ before a new token $(i+1)$ in the other phase is provided to the input of register n, the new capturing process will start as soon as the new token is available at the input of register n.

2. If the preceding stage $(n-1)$ provides a new token $(i+1)$ to the input of register n before the subsequent stage $(n+1)$ has acknowledged the last token i, the new capturing process at register n will start as soon as the acknowledge signal (*Done* output of register $(n+1)$) is received on the *Pass* input of register n.

In case of a stuck-at fault either the new token is inconsistent and will thus not be processed, or the acknowledge signal has the wrong state. In both cases the capturing process will not be started. Transient faults causing toggling signals could even lead to the loss of tokens [28], but transient effects are out of scope for this thesis and thus not considered in the following definitions.

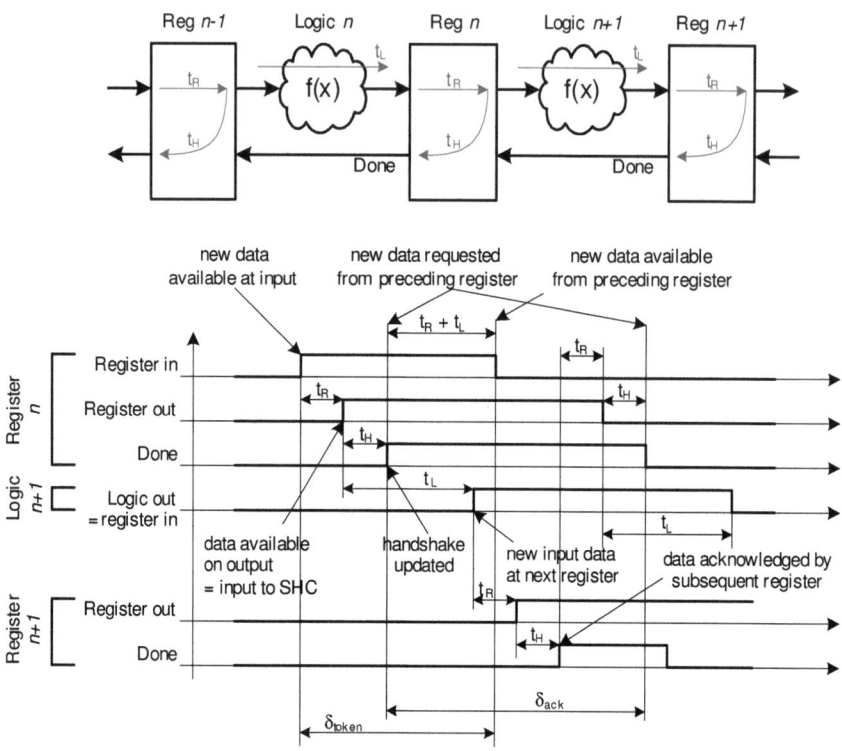

Figure 2.7: Timing Parameters

The time δ_{ack} is defined as the duration between two subsequent toggles of the *Done* signal of the same register, and δ_{token} is the duration between two subsequent tokens (in alternating phases) being available at the same register input. In order to guarantee a correct capturing two constraints must be fulfilled:

Constraint 1 *The register n must not receive the acknowledge from the subsequent stage $(n+1)$ for capturing token i before register n has captured the token i and has acknowledged this to the preceding stage $(n-1)$: $\delta_{ack}(n+1) > t_R(n) + t_H(n)$*

Constraint 2 *Any register must not receive a new token at its input before the last token has been captured and acknowledged: $\delta_{token}(n) > t_R(n) + t_H(n)$*

If any of these constraints is violated, the register might either not capture the data or capture wrong data. In a standard pipeline as shown in Figure 2.7 δ_{ack} is determined by the time the preceding stage $(n-1)$ needs to provide new data after stage n has issued the *Done*

2.4. Four-State-Logic

signal (return path), and by the time the subsequent stage $(n+1)$ needs to capture the last token and assign the *Done* signal correctly (forward path).

Equation 1 $\delta_{ack} = max(\delta_{ack,min}(return); \delta_{ack,min}(forward))$

Equation 2 $\delta_{ack,min}(return)(n) = t_R(n-1) + t_L(n-1) + t_R(n) + t_H(n)$

Equation 3 $\delta_{ack,min}(forward)(n) = t_L(n) + t_R(n+1) + t_H(n+1) + t_R(n)$

Note: Equation 2 describes the minimum time, assuming that register $(n-1)$ receives valid inputs fast enough.

The duration δ_{token} is defined by the time the register n needs to capture the token i and by the time the new token $(i+1)$ needs to propagate from the preceding stage to the input of register n after the *Done* signal was asserted.

Equation 4 $\delta_{token,min}(n) = t_R(n) + t_H(n) + t_R(n-1) + t_L(n-1)$

In a practical implementation constraint 2 is always fulfilled in the fault-free case because register $(n-1)$ will not become transparent and issue a new token $(i+1)$ before register n has acknowledged the capturing of token i. Constraint 1 would only be violated if the handshake path of register n would be slower than the sum of data and acknowledge path of stage $(n+1)$, which can easily be avoided by careful routing.

2.4.5 Faults in Asynchronous Circuits

Asynchronous circuits behave different to synchronous circuits in many points. They are insensitive to timing variations and are therefore robust against many radiation effects that affect timing [60]. However, SEUs can cause deadlocks or other types of erroneous behavior. Traditional methods such as TMR cannot easily be applied to asynchronous circuits, as a fault in the asynchronous logic could prevent the result to appear at the output and thus block the comparison permanently [102, 101].

A detailed investigation of the effect of various fault types on asynchronous circuits is presented in [76, 80]. Methods to harden asynchronous circuits against transient faults/errors are presented in [60, 61, 88, 89, 90, 108] but are not considered any further in this chapter as the thesis focusses on permanent errors. In [103] and [49] various methods of testing asynchronous circuits are presented.

With FSL registers and combinational logic designed as described above, only consistent data will propagate through a pipeline. Permanent faults at the handshake lines between elements or at the data inputs/outputs will either directly stop the handshake process or cause inconsistent data and indirectly lead to a deadlock. Any permanent fault inside the elements (e.g. register control logic) will show an effect on the external interfaces (e.g. inconsistent data, wrong handshake signal, etc.) and thus can be covered by an appropriate fault model on pipeline level (for details refer to chapter 4).

However, the anticipation that only consistent data is processed in an FSL pipeline is only true on boolean level, i.e. in the fault-free case. As soon as a permanent fault exists in an FSL circuit it is not hazard-free any more. As will be shown below, this can lead to the case that consistent, but wrong data moves through the pipeline.

As an example, such a case is given for the FSL AND gate described in a previous section. A fault at the b-rail of input E2 shall be assumed. The applied tokens will thus be changed to inconsistent tokens if the affected rail would have a different logic state than appears due to the fault.

Figure 2.8 below presents the Karnaugh map for the Set-signal of the latch defining the a-rail of the output. Table 2.2 lists a sequence of input tokens to the AND gate, the resulting input pattern due to the fault and the effect if the transition causes a hazard. It can be seen that due to the fault there exist transitions which can cause hazards and generate consistent data values, which are, however, wrong.

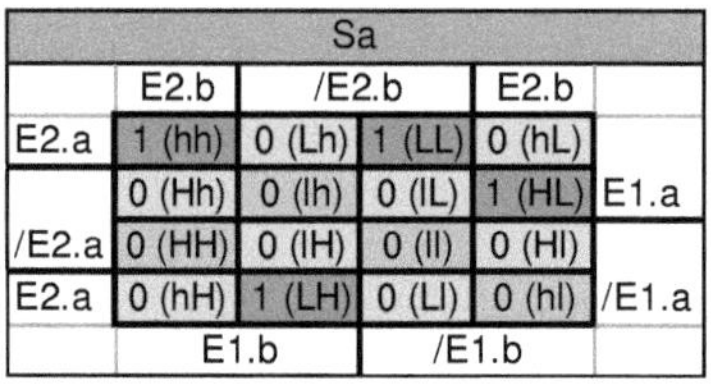

	Sa				
	E2.b	/E2.b	E2.b		
E2.a	1 (hh)	0 (Lh)	1 (LL)	0 (hL)	
	0 (Hh)	0 (lh)	0 (IL)	1 (HL)	E1.a
/E2.a	0 (HH)	0 (lH)	0 (ll)	0 (HI)	
E2.a	0 (hH)	1 (LH)	0 (Ll)	0 (hl)	/E1.a
	E1.b		/E1.b		

Figure 2.8: Karnough Chart of Sa Signal for 2-Input FSL AND Gate

Applied		Faulty Token		Transition	Output	Expected	Status
ll	(0000)	⇒ lH	(0010)				
LL	(0101)	⇒ Lh	(0111)	via LH	L	L	O.K.
hl	(1100)	⇒ hH	(1110)	via hh	h	l	WRONG
LL	(0101)	⇒ Lh	(0111)	via LH	L	L	O.K.
hl	(1100)	⇒ hH	(1110)	via hh	h	l	WRONG

Table 2.2: Exemplary Token Sequence and Effect for Faulty FSL Logic Input

This issue has not been treated within this thesis, as it is an inherent problem of FSL logic and needs to be handled on higher level. A detailed investigation of such timing dependent effects and their probability of occurrence can be found in [28].

For the simulations and experiments the combinational logic has been chosen appropriately so that the described effect does not occur. Details are described in chapter 5.

Chapter 3

State of the Art of Circuit Reconfiguration

3.1 Introduction to Autonomous Self-Repair

So far, achieving high hardware reliability in critical applications (e.g. for space missions) was only possible by combining the three measures (i) reducing component stress, (ii) increasing component quality and (iii) implementing redundancy (see section 2.2.2). It is quite obvious that traditional repair, e.g. exchanging a defective component, cannot be performed as soon as the affected electronics is in space. Consequently, since the fault density is increasing [1], the traditional methods of reducing component stress and screening the components might not be sufficient any more.

Redundant systems are typically designed to be tolerant against a single fault. After the first fault the redundant part can take over, but a second fault occurring in the redundant part will lead to a system failure. One possible way to improve the tolerance also against multiple faults and to increase the reliability is to repair the faulty part to bring it back to operation. If the circuit/system cannot be repaired by exchanging hardware elements, the circuit must be reconfigured so that defective resources are bypassed and replaced by working ones. Systems that can perform such repair-actions autonomously are called *self-healing* systems.

The term *self-healing* and its differentiation to self-repair and fault-tolerance is extensively discussed [71] and still not clearly defined. One basic identification is that for fault tolerance the aim is to keep the system at 100% functionality, while self-healing systems allow to operate at less than 100% after the healing procedure [111].

The following definitions of self-healing were published:

- *Tosi* [126] describes the need for *self-management* to handle the increasing complexity of computing systems. Due to the increasing state-space of fault combinations not all effects can be foreseen. As a consequence, systems that can adopt to new situations and conditions as they arise are needed. *Self-healing* is one important category of self-management and is defined as *"the system ability to examine, find, diagnose and react to system malfunctions"*.

3. State of the Art of Circuit Reconfiguration

- According to *Saha* [112], who describes a software-based self-healing system, *"self-healing deals with imprecise specification, uncontrolled environment and reconfiguration of system according to its dynamics"*.

- *Gericota* [36] describes self-reconfiguration (of FPGAs) as a *"method to give the currently configured functions the control of (re-)configuring areas of the same FPGA"*.

- *Rodosek* et.al [111]: *"A system is showing the self-healing characteristic if it is able to monitor and heal itself from the inside, which requires the ability of this system to decide about and perform recovery actions to return itself to a behavior conforming to its initial specification, especially without external interference"*.

- *Laster* et.al [113] describes self-healing as a closed-loop cycle, where the processes "monitoring", "error detection and diagnosis", "analysis and selection of a repair operation" and "execute repair and operation (self-repair)" are continuously followed.

Self-repair is already state-of-the-art in regular circuit structures such as memories [54]. In irregular circuits self-repair is a complex task, but gains more and more interest as redundant resources can be used in a much finer granularity and distributed redundancy and repair (distributed self-healing) can significantly increase the circuit's lifetime [105].

Within this thesis the term "self-repair" is used to describe the process of bringing the circuit back into operation. "Self-healing" focusses on the autonomy and includes all actions which lead to the repair process. A self-healing system thus must be able to perform autonomously the following tasks:

1. fault detection: detect the fault, either offline or online

2. fault diagnosis: identify the faulty unit/element in the circuit/system

3. fault isolation: remove the faulty element from the operational circuit, e.g. by switches

4. redundancy allocation: replace the defective parts by redundant elements

5. repair validation: test/validate the performed repair

In scope of this thesis the self-healing procedure (including the circuit reconfiguration) shall be performed during runtime, thus not only the logic functionality and the interconnections, but also the internal states have to be restored. The following section assesses the possibilities for circuit reconfiguration in integrated digital circuits. A summary and conclusion about the suitability for self-healing is presented at the end of this chapter.

3.2 Circuit Reconfiguration

3.2.1 Runtime Reconfiguration with FPGAs

This section does not describe a dedicated method for circuit reconfiguration, but gives a general overview about FPGAs and in particular about those that offer (runtime-) reconfiguration capabilities. This feature can be used in several approaches described in the subsequent paragraphs.

Field Programmable Gate Arrays (FPGA) are integrated circuits that consist of configurable logic blocks (CLBs), vertical and horizontal routing paths, and programmable interconnections (switch boxes). Figure 3.1 [18] shows the basic architecture of an FPGA. The actual function of the circuit as well as the interconnections are programmed by the user. Depending on the type of FPGA the configuration is either stored in an internal SRAM (*SRAM-based FPGA*, e.g. [56, 10]), which can also be updated, or by burning fuses (*anti-fuse FPGA*), which results in one-time-programmable devices [5].

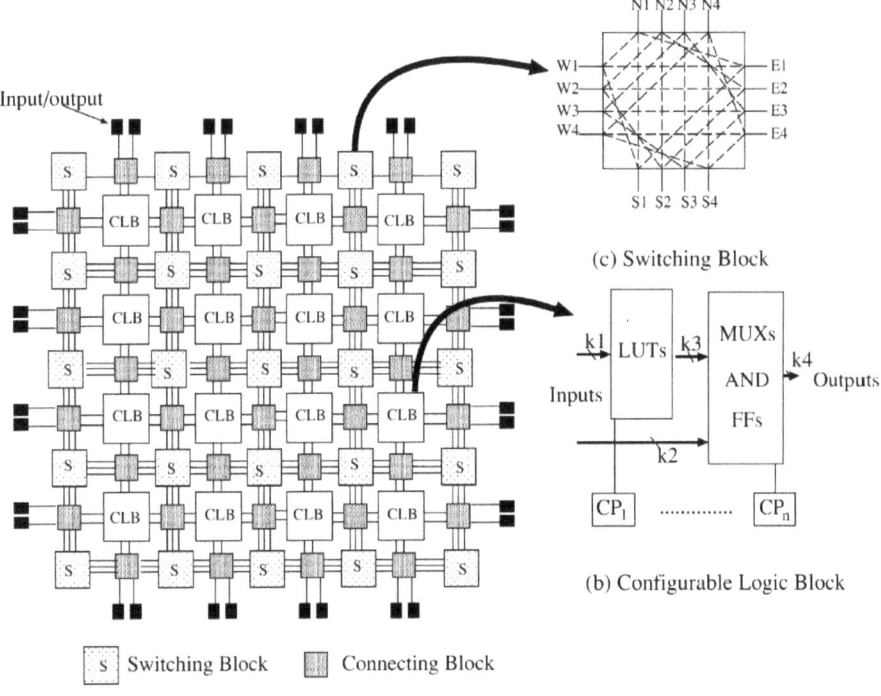

Figure 3.1: Basic FPGA Architecture [18]

In the past FPGAs were mostly used for prototyping and low volume applications. Increased performance, memory size and flexibility of modern FPGAs, as well as lower costs, make them now an attractive alternative to ASICs.

A rather new area of FPGA applications is *reconfigurable computing*, which allows to share hardware between different applications, to adapt hardware algorithms, increase the resource utilization, and allows to upgrade the hardware remotely [66]. Possible applications in this scope are e.g. reconfigurable accelerator processors in supercomputers [55], where algorithms are out-sourced into hardware to speed up calculations. The same component can be used for other calculations at a later point in time.

Although radiation-hardened FPGAs are available, the SRAM is susceptible to SEUs [42], and so there is still great scepticism in using SRAM-based FPGAs in e.g. space applications. Currently, mainly anti-fuse types such as the Actel Axcelerator series [7] are used for high reliability applications being operated in extreme environments.

The Xilinx Virtex series is SRAM-based and can be reconfigured during runtime, called *dynamic* or *runtime reconfiguration*. As only this series is suitable for self-repair during circuit operation, the following explanations refer to the Virtex series. In particular, Virtex FPGAs also can be *partially reconfigured*, which allows to update specific parts of the circuit while the remaining circuit is operating.

The circuit synthesis and place&route process requires powerful processors and may take minutes to hours, and is thus usually not feasible to be performed in an embedded system during operation. Various approaches have been proposed to speed up this task and make it possible to be performed online [82, 84, 85, 6]. The only tool for runtime routing provided from a FPGA manufacturer is JBits and the extension JRTR [43, 83], which is, however, restricted to the Virtex-II series.

The Virtex series offers different interfaces that allow external (serial, SelectMap, JTAG) as well as internal configuration (ICAP; controlled from the application), which differ in the data width and the clock speed [133].

The smallest unit which can be read or written to/from Virtex FPGAs is a *frame*. The size of a frame depends on the FPGA type and limits the granularity for any applied configuration.

3.2.2 Dynamic Rotation and Free for Test

The *Dynamic Rotation And Free for Test* (DRAFT) method [31, 32] is a non-intrusive approach for on-line concurrent testing, detecting and avoiding permanent faults, and correcting errors due to transient faults by utilizing active, respectively dynamic replication. This method is suitable for FPGAs, in particular the Xilinx XCV200 was used. All resources of the FPGA are continuously tested on CLB level through the whole system life time. The CLB to be checked is copied to a free CLB (Figure 3.2, [34]) and then tested. If an error is detected, the location is marked as erroneous and avoided for future use.

The DRAFT method performs an online reconfiguration by means of a two-phase replication (Figure 3.3). In the first phase the configuration of the CLB is replicated and the inputs of the two CLBs are connected in parallel. After at least one clock cycle both CLBs have the same internal state. In the second phase the outputs of the replicated CLB are connected to the circuit. For transferring the internal state three cases have to be considered: synchronous free-running clock circuits, gated-clock circuits and asynchronous circuits.

3.2. Circuit Reconfiguration

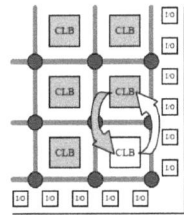

Figure 3.2: CLB Replication and Rotation of Free Resources [34]

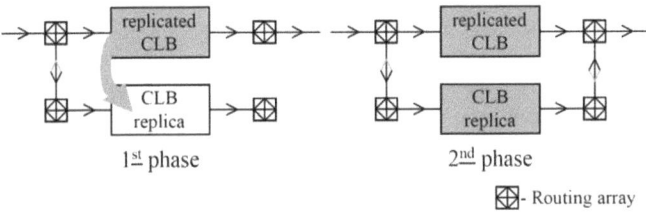

Figure 3.3: Two Phase Replication [31]

For synchronous free-running clocks the two-phase replication will ensure the correct state in the replica. In gated-clock circuits a replication aid-block is used, as shown in Figure 3.4.

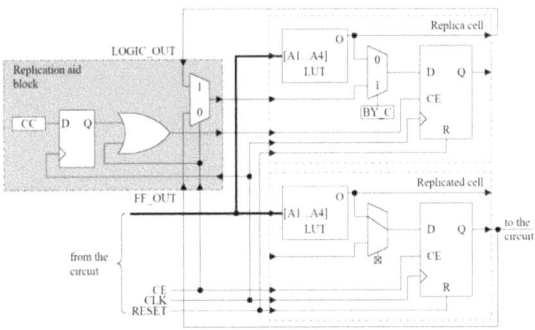

Figure 3.4: Replication Scheme for Gated-Clock Circuits [31]

The replication aid-block establishes a temporary transfer path between the replica and the replicated cell. If only free routing resources are used this does not influence the functionality of the circuit. With the aid-block the state-information can be transferred to the replica cell. The same method is suitable for asynchronous circuits or circuits with multiple clocks. In any case a clock is needed as trigger for transferring the internal state.

Synchronous free-running clock circuits do not copy the internal state, but acquire the values as the replicated cell does. Therefore the replica cell shows the correct value afterwards, even if there is an error in the original cell. However, in gated-clock or asynchronous circuits, the state of the original cell is transferred to the replica cell with the aid-block. If the value of the replicated cell is erroneous, the fault is propagated to the replica cell. For detection of these errors a higher level routine has to be applied.

The number of faults that can be handled depends on the number of spare CLBs that can be used for testing and replacement. At least one CLB in the FPGA must be free, which is used to test another CLB. For the different phases of the replication process several configuration files are needed, which need to be pre-defined during the design phase. The configuration is loaded via the FPGA configuration interface.

If LUTs are configured as RAM, the replication process is not possible as memory coherence problems can occur. The system would have to be stopped during replication to be able to cope with this problem.

A dedicated controller is needed, which handles the regular testing and replacement in case of faults. Interconnect faults are not considered in this approach.

3.2.3 Fine-Grained Self-Healing Hardware

The approach introduced in [130] describes a fine-grained method for implementing self-healing hardware. Small amounts of reconfigurable hardware are used to implement redundancy for multiple *cones*. A cone is defined as a combinational logic block that has several input signals and one output. Any component can be split into a set of cones, and look-up tables (LUTs) can implement the functionality of any cone with a defined maximum number of inputs. The size of the LUT depends only on the number of inputs and not on the complexity of the implemented function. Figure 3.5 shows the principle of this method.

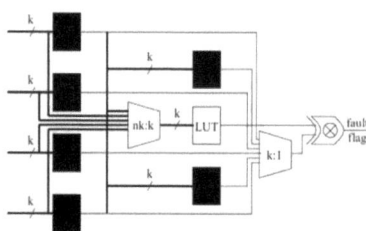

Figure 3.5: Cone-Level Fault Detection and Diagnosis [130]

First, the LUT is configured to implement the functionality of the cone to be tested. Then the nk:k multiplexer selects one cone (respectively its inputs) and another multiplexer selects the corresponding output. The cone output and the multiplexer output are compared with the XOR gate, and an error signal (fault flag) is generated if they are different. If the cone currently tested is detected to be faulty, the LUT can take over its function and the error is hidden (self-healing).

3.2. Circuit Reconfiguration

The timing overhead is small since only a low number of bits need to be reconfigured to implement the cone function in the LUT and the testing is performed in parallel to the remaining functionality.

The cones have to be extracted during the design time, as well as the LUT SRAM configuration, placement and cone routing. All bit configurations have to be pre-generated and stored in either on-chip or off-chip memory.

The fine-grained self-healing hardware approach can only handle combinational circuits. A dedicated controller is needed, which handles the regular testing and replacement in case of faults. Interconnect faults are not addressed in this approach.

3.2.4 Dynamic Reconfiguration using Atomic Fault Tolerant Blocks

The key elements of the method described in [74] are *tiles* and *atomic fault tolerant blocks* (AFTB). A tile consists of three elements: (i) a set of CLBs and interconnect resources, (ii) a net-list which must be placed on those CLBs and (iii) an interface specification of how the tile connects to adjacent tiles. The AFTB is one instance of a tile and contains at least one spare CLB that serves to cover the faulty CLB. The circuit is partitioned into tiles which all have the same interface to the surrounding areas of the design. The tile contains the logic circuit implemented in CLBs and at least one free CLB. For this circuit functionality several placements within the tile are possible so that the free CLB moves to different locations (Figure 3.6). With this approach the active circuit can be moved away from faulty resources.

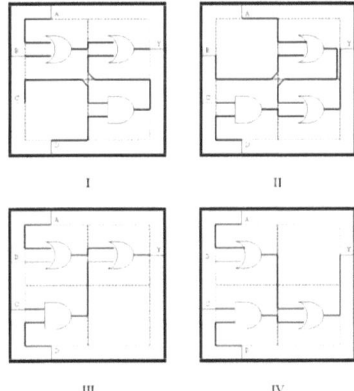

Figure 3.6: Tile-Based Design [74]

This method can handle multiple faults in the circuit, but only one faulty CLB per tile. Local interconnect faults can be handled in the same manner. If an interconnect line is defective, a routing solution not using this line is configured. Faults in global interconnections require different handling since much of the interconnect crosses tile boundaries. Global lines are used as backup to replace defective interconnect lines [75].

3. State of the Art of Circuit Reconfiguration

Although the basic structure of course remains the same, the actual implementation of the algorithm depends on the FPGA architecture. The partitioning into tiles as well as the different routings have to be generated during design time. These configurations are stored in memory and are used for reconfiguration depending on the location of the faulty cell. The repair process can be performed during runtime if the FPGA supports runtime reconfiguration. The timing effort is then rather low, as partial reconfiguration can be used.

This approach does not include error detection and fault localization.

3.2.5 Column-Based Precompiled Configuration

In this method the FPGA, respectively the design, is partitioned into columns which can be shifted, so that faulty resources are moved to unused areas in the FPGA [50]. An example for the overlapping precompiled reconfiguration scheme is given in Figure 3.7. The base configuration of the design is mapped in four consecutive CLB columns. The rightmost column is reserved as backup column for alternative configurations. In the right picture column three is faulty. All columns to the right of the faulty column are shifted rightwards using the spare column so that the faulty resource is then unused. Only inter-region signals need to be rerouted. All signals within the column remain the same since the CLB columns in the FPGA contain the same programmable logic and routing resources.

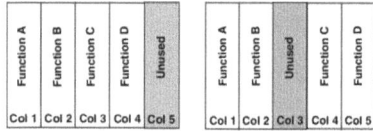

Figure 3.7: Column-Based Precompiled Configuration [50]

Another possibility is the non-overlapping scheme, which can be applied if the circuit is small enough to fit within half of the FPGA. It is then not necessary to precisely find out the location of the faulty column. To minimize fault location overhead, all possible configurations can be tried until a working solution is found.

For error correction the *dual FPGA architecture* is proposed. Each of the two FPGAs in figure 3.8 implements a soft-microprocessor of which one is able to reconfigure the other FPGA and vice versa.

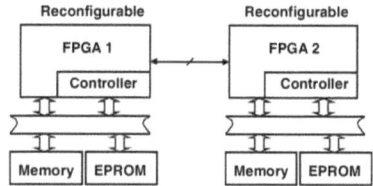

Figure 3.8: Dual FPGA Architecture [50]

3.2. Circuit Reconfiguration

It is important that the communication channels between the two FPGAs are working correctly in all situations. Since this is a vital part of this method, the communication channel has to be made fault tolerant [87]. Furthermore, the microcontrollers have to be able to recover from faults. Appropriate techniques such as TMR inside the FPGA have to be considered.

The partitioning into columns has to be considered during the design phase, and the different configurations have to be pre-compiled.

By only shifting the columns, similarities of the configurations can be achieved. This allows to use data compression efficiently and thus to reduce storage overhead. However, a lot of functional resources might be lost in a replaced column and so the number of configurations soon reaches a limit if multiple faults are considered.

In the overlapping scheme this method is able to tolerate m faulty columns where m is the the number of spare columns reserved in the FPGA. For the non-overlapping scheme the base configuration has to fit into $1/(m+1)$ columns to tolerate m faulty columns. The base configuration should take care that most of the signals flow in vertical direction, since the maximum number of horizontal routes is limited.

If the LUTs are used for RAMs, memory coherence problems can occur if write attempts are performed between readback and writeback. This issue is described in detail in [51] and [52].

The internal states are not transferred to the new column, thus this method is not transparent to the application.

3.2.6 Roving STARS

This method is based on *roving self-testing areas (STARs)* [21]. The STAR is a small roving portion of the FPGA which is tested off-line while the other parts remain on-line and continue normal system operation. This approach therefore integrates on-line test, diagnosis and fault tolerance within the same framework.

As can be seen in Figure 3.9 there are one horizontal and one vertical STAR. The horizontal STAR moves up and down, while the vertical STAR moves across the FPGA and back again. Both STARs are required to test the interconnect resources. Since the STAR is a currently unused area in the FPGA it can be tested without affecting the remaining functionality. After testing has been completed, the STAR moves to a new location. This is performed by partial reconfiguration, i.e. by copying a part of the working system into a previous STAR. Then the system clock is stopped for a few cycles to transfer the state to the copied region. After that, the clock is restarted and the new STAR can be tested. Apart from that, no timing overhead occurs.

Any fault detected during testing is located in a STAR and does therefore not affect the system. The reconfiguration is controlled by an external controller, usually an embedded microprocessor. This controller has to keep track of the status of the FPGA, in particular in which section the different functions are currently implemented and where the faulty resources are, to avoid their use in future configurations. In [21] it is assumed that the controller also performs the test, diagnosis and fault-tolerance functions including their associated reconfigurations and is therefore called *test and reconfiguration controller (TREC)*. The TREC controls all actions via the boundary scan interface.

Testing is performed with an integrated test pattern generator which applies pseudo-random patterns to two identically configured programmable logic blocks (within a STAR) under test.

3. State of the Art of Circuit Reconfiguration

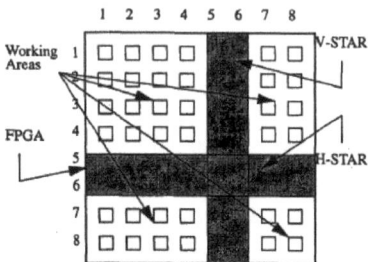

Figure 3.9: Roving STARs [21]

The outputs are directly compared, thus avoiding overhead of storing expected output patterns. The logic blocks are continuously configured in all possible modes of operation, such as adder, flip-flops, RAM, LUTS. This approach allows to define *partial usable blocks*, which means that in a programmable logic block e.g. a flip-flop is defective but the combinational resources can though be used. Before re-mapping the area to a working area again, it is reconfigured around faulty resources to avoid using the defective parts.

In level two the FPGA has to be reconfigured to avoid usage of the faulty resources. For single faults, for each system logic cell function a partial configuration can be pre-compiled and stored in the memory. For multiple faults pre-compiled configurations are not feasible any more so processing has to be done online by the TREC. Since the faulty resource is covered by a STAR, the system functionality is not affected even if the online processing takes a long time.

There must be enough spare area available in the FPGA to have one horizontal and one vertical STAR.

3.2.7 Node Covering Technique

The method of node covering [44, 45] can be applied to SRAM-based FPGAs and allows each cell[1] to replace its neighbor in the same row. To each cell in the FPGA a cover cell is assigned which can be reconfigured to take over the functionality of the other cell in case it becomes faulty. Applying the node covering method allows to tolerate one fault in each row or column. Since all cells in an FPGA are equal, all cells can cover each other with respect to functionality, and the configuration can simply be transposed to the other cell without the need for re-routing. The difficult issue is to establish the correct connectivity, which requires additional wiring segments.

In the example in Figure 3.10 the rightmost cells are spares. The cell-to-channel interconnections belong to the configuration data and do not need to be modified since the complete cell configuration is transposed. However, the channel segment interconnections require *cover segments*, bordering the cover cell. Such segments may either be already in the correct position (e.g. channel segment between A and B in figure 3.10) or *reserved segments* have to be implemented. Reserved segments are spare routing resources for the basic configuration but are

---
[1]A cell is a programmable combinational function with an optional output register, i.e. a CLB.

3.2. Circuit Reconfiguration

needed if the covering procedure is performed. For example, the connection between B and C in Figure 3.10 can only be established via a reserved segment. Covering vertical segments may require up to two reserved segments [44].

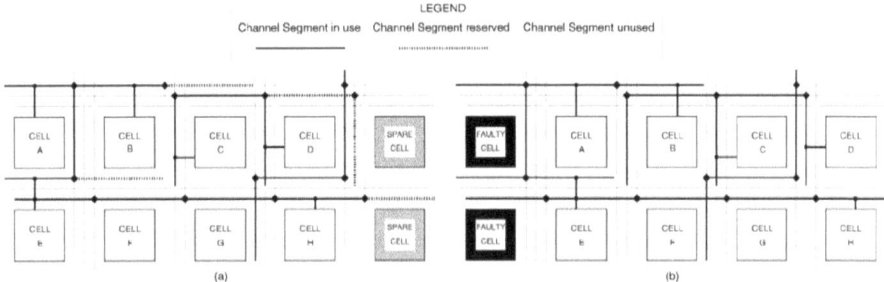

Figure 3.10: Node Covering Methodology [45]

Due to the additional segments (reserved segments) the propagation delay will increase. In the fault-free case, the reserved segments need not be connected. This requires two different configurations for each row, resulting in a 30% overhead of the configuration data [44]. Alternatively, a more sophisticated method can be applied which identifies a re-configured cell and deduces from that whether the reserved segment shall be connected or not.

State recovery is not covered by the method, and an external error detection and reconfiguration controller is needed. This approach can handle one fault in each row or column, but it is not intended to be used during operation of the application.

3.2.8 Method of Shifting Configuration Data

Due to the homogeneous structure of FPGAs it is possible to achieve fault tolerance by shifting the configuration data [19] to move the circuit to another region avoiding the usage of the defective resource. The approach can be applied to SRAM-based FPGAs. With this method both defects in CLBs as well as in interconnect resources can be tolerated. The prerequisite for this method is to have spare CLBs regularly distributed among the whole FPGA as well as at the borders. If an error is detected, the user data is shifted to the nearest spare CLB. For the allocation of the spare CLBs two methods are described: king-shifting and horse-shifting allocation.

If a defect is detected, the configuration is shifted to the corresponding direction, depending on the defect location. The defective CLB is unused after the shifting. Since all unit elements are shifted, the interconnections between the CLBs do not change. An example for a shift is given in Figure 3.11.

The king-shifting allocation requires fewer spares since only every nine CLBs one spare is needed, while the horse-shift method requires one spare every five CLBs. Since the user data remains the same after shifting, there is no additional delay introduced.

This method is able to tolerate faults in CLBs and in connection blocks which are connected directly to the CLBs. For interconnect resources this method also can be applied, but with

3. State of the Art of Circuit Reconfiguration

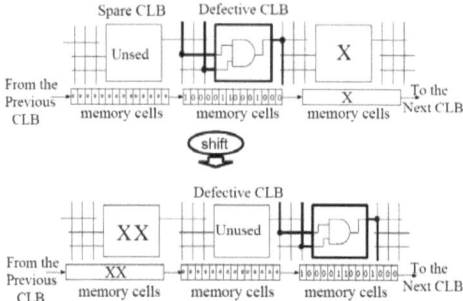

Figure 3.11: Example of Shifting the Configuration Data [19]

some restrictions. It will fail if all eight elements around the faulty CLB are used by the user application.

The reconfiguration is performed with the whole application, i.e. the complete configuration is shifted. The number of faults that can be tolerated is thus quite limited.

This method is not intended to be used during operation of the application.

3.2.9 Nature-Inspired Methods

The *embryonic electronics* architecture, called *embryonics* [124], describes an approach for self-repair in integrated circuits using both S/W and H/W redundancy. The structure of integrated circuits is mapped to molecules, cells and genomes. An organism is able to detect faults and repair them without any centralized controller by the two properties self-repair and self-replication. In Figure 3.12 it is shown how the embryonic approach can be implemented on an electronic circuit.

The circuit (organism) is constructed from universal cells, where each cell is a small processor which executes its gene program. For this approach a new kind of FPGA is proposed. The principle of the algorithm has been successfully implemented on a prototype system.

The benefit of the embryonic approach for memories has been investigated in [106] with explicit focus on soft error tolerance in space applications. It is stated that embryonics alone does not solve the problems, but it is advantageous if it is applied in addition to traditional error correction techniques. Both approaches [124, 106] are only described theoretically, but no practical measurements or hardware implementations are available.

Genetic algorithms that can autonomously evolve a new working circuit have been successfully implemented with combinational logic [96]. The repeatability of the results is, however, considered rather unpredictable. *Vigander* [129] applied a genetic algorithm to a traditional voting system to improve the long term fault tolerance, but only simulations have been performed.

In [104] another method inspired by the human immune system is described. Here, the resources are split into identical functional cells. The function of each cell is defined by the genetic code. The functional cells are surrounded by spare cells which can clone a functional

3.2. Circuit Reconfiguration

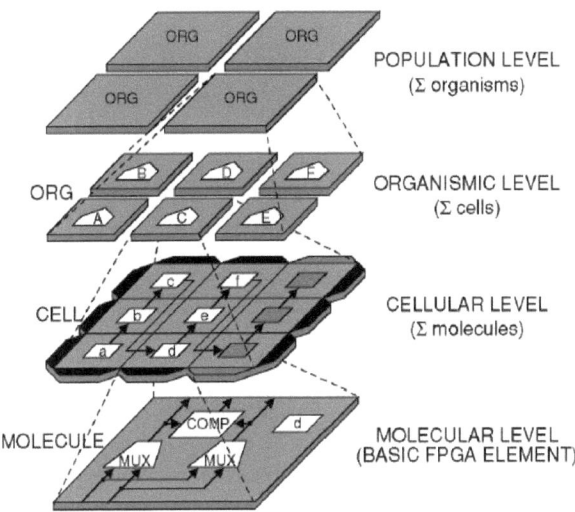

Figure 3.12: Embryonics Landscape [124]

cell. This approach, however, is only described theoretically and no details how to implement this method in a real circuit are given.

For all mentioned approaches internal state recovery is not addressed.

3.2.10 Self-Repair using Re-Configurable Logic Blocks (RLBs)

The approach published by Vierhaus et.al. [37, 69, 68] is based on reconfigurable logic blocks (RLBs) that contain a particular number of gates, of which (at least) one can be selected to replace any of the other ones. Each gate can be completely isolated by means of switches at its input and output (Figure 3.13).

The RLBs can contain basic gates (AND, OR, etc.) or even complex circuits (e.g. an adder). Depending on the number of elements, the RLB has several states. In each state one or more of the functional elements are replaced by spare elements.

This approach is implemented on transistor level. The overhead is quite high for basic gates (e.g. 230% for a 2-input NAND gate), but can be significantly reduced if the blocks become more complex (e.g. 38% for an 8-bit ALU), as the overhead is mainly introduced by the switching elements. It is mentioned that sequential logic as basic block is not feasible due to the complex wiring. Furthermore, the overhead due to the test pattern generation and diagnosis is not included in the numbers. Larger blocks might become even more efficient, but reduce the multiple fault tolerance. Although the wiring between gates is not explicitly covered by the repair scheme, for larger blocks the majority of the total wiring is within the block and so it is likely that also faults in the wiring/interconnects can be repaired.

3. State of the Art of Circuit Reconfiguration

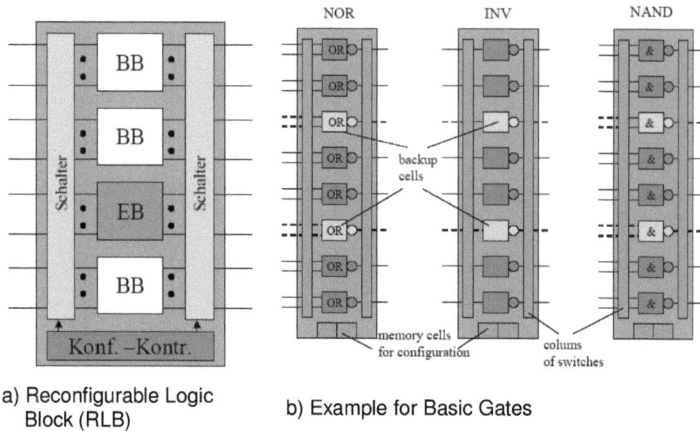

a) Reconfigurable Logic Block (RLB)

b) Example for Basic Gates

Figure 3.13: Reconfigurable Logic Blocks (RLBs) [37, 69]

As simple and efficient solutions are only possible for regular structures (e.g. memories), an approach to extract regular units from irregular netlists was investigated. This allows to reach reasonable and efficient sizes for basic RLBs [37] while still maintaining feasible fault densities.

The reconfiguration principle is shown in Figure 3.14. After a test pattern has been applied, the output is compared to a reference value, both being provided e.g. via scan chains. If the results are different, there obviously exists a fault in at least one element. A 2-bit counter is triggered by this information and changes the configuration of the RLBs. Then the same pattern is checked again. This sequence is repeated until there is either no difference detected, or all configurations have been tried out. If no solution can be found, an error flag is set to indicate this status to the higher level system.

The reconfiguration can be controlled from a central instance or from local controllers. Although the central approach is more efficient in terms of resources, the high amount of signals that need to be routed to the controller are considered as potential problem for transient faults. This approach is also proposed to be used for thermal de-stressing of resources.

The approach using reconfigurable logic blocks determines the required reconfiguration pattern during runtime (built-in self repair), however, for the reconfiguration process the application has to be stopped and the circuit has to enter a dedicated test mode. It is proposed that this is done after power up to configure a correctly working circuit. The gain in reliability is not quantified for the time being, but several ideas for further improvements are proposed, such as implementing the repair circuitry more robust by using thicker oxides for transistors. As in all redundant systems, also here the switching elements are the bottleneck and need to be highly reliable in order to increase the overall reliability.

3.2. Circuit Reconfiguration

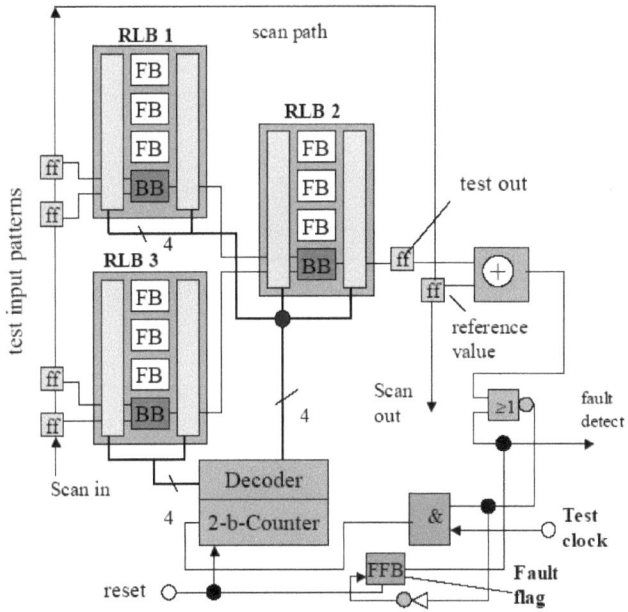

Figure 3.14: Concept of Self-Repair with RLBs [37]

3.2.11 Self-Healing Asynchronous Arrays

The method presented in [102] is dedicated to asynchronous logic. Linear arrays are defined as basic building blocks, as most circuits can be designed from such arrays.

Due to the multi-rail encoded data representation these circuits provide self-checking capability, which can be extended to fail-stop behavior with low effort [100]. The deadlock is then detected by a watchdog, and a reconfiguration procedure can be started. The principle is shown in Figure 3.15.

The method is applied on transistor level. In particular a *precharge half buffer (PCHB)* has been chosen as circuit template, which can be used to construct almost any pipelined QDI logic. The deadlock detector (watchdog) waits for the next valid protocol state to occur by checking the handshake activity. If this does not happen for a dedicated time, a deadlock is assumed. The delay is implemented using a delay line which allows delays in the order of milliseconds. Two methods for online reconfiguration are proposed: (i) locate faults and use a workable configuration or (ii) try all possible solutions until a workable one is found. Although the first method achieves a faster fault recovery time, the required logic can be quite large which decreases the overall reliability. The second method is slower but also needs less hardware overhead. The slower procedure does not influence the overall performance significantly since faults are expected to be rare. The second method is thus proposed to be used and has the advantage, that all possible configurations would be re-tried for subsequent errors and so also

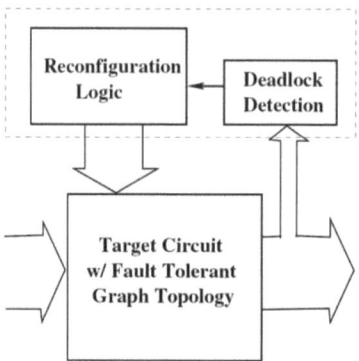

Figure 3.15: Block Diagram of a Reconfigurable Self-Healing Asynchronous Array [102]

transient effects can be resolved. After a reconfiguration the circuit needs to be reset and re-started from the last system entry point.

A fault-tolerant asynchronous adder has been successfully implemented based on this approach [101].

The logic for deadlock detection is not described in detail, but it is said that this part should be designed conservatively to achieve high robustness. Furthermore, the reconfiguration logic is proposed to be implemented as synchronous logic, as it requires less hardware resources.

3.3 Comparison and Conclusion

Due to the increasing fault density in integrated circuits, the traditional means for fault avoidance might not be sufficient any more to reach the reliability goal of an application. Self-repair seems to be a promising way to implement fine granular fault tolerance that can be handled even during runtime and transparent to the application.

Triple-modular redundancy is one of the most commonly known approaches for fault tolerance. Basically, TMR is mainly suitable for transient fault tolerance. Considering also permanent faults, the reliability soon decreases for longer mission times [115]. Furthermore, TMR implies high costs in terms of resources (factor > 3) and power consumption. In principle all reconfiguration methods can be used to implement a self-healing TMR architecture, which then might even provide long-time fault tolerance against permanent faults.

Most of the approaches require dedicated fault detection, diagnosis and localization techniques before a fault situation can be handled. While in clocked circuits it is simple to stop the execution for performing recovery actions, in asynchronous circuits particular means have to be considered, such as forcing a deadlock. However, asynchronous circuits inherently tend to stop their execution in case of permanent faults [76] and are thus advantageous in this sense.

Asynchronous circuits require storage elements even for combinational functions, thus the method described in [130] is not suitable, as it only covers LUT functions. Although the other methods basically seem to be usable for asynchronous circuits, they have not been investigated for this kind of circuits.

3.3. Comparison and Conclusion

Logic reconfiguration in FPGAs is quite handy if the FPGA supports partial and/or runtime reconfiguration, however, the FPGA configuration interface is a single-point-of-failure, and is thus considered as a critical part for the methods that depend on this interface. Furthermore, the FPGA resources can be fragmented and thus the probability for finding a new configuration might be reduced [33].

All known methods of runtime reconfiguration for permanent error recovery in FPGAs implement spare resources of different granularity, which are used to bypass defective circuit parts. Some methods require spare combinational logic blocks (CLBs) to be regularly distributed across the FPGA [19] or just at particular positions [44, 74]. Other methods use even large spare areas in an FPGA that occupy whole columns [50]. The size and distribution of these spare resources also defines the granularity of a possible reconfiguration.

The majority of the FPGA-based approaches [31, 130, 74, 50, 21, 44, 19] requires predefined circuit configurations which have to be stored in (large) memories. In order to achieve a gain in reliability, these memories also have to be highly fault tolerant. Additionally, the above mentioned methods require a dedicated controller that performs all necessary tasks like masking of defective resources, transferring internal states, copying CLB configurations, etc.. Online circuit synthesis is complex and very time consuming [84], and is not considered possible to be performed during runtime in an embedded dependable system.

Genetic algorithms can autonomously evolve a new working circuit. Although successful implementations for combinational logic were presented [96], the result seems to be rather unpredictable and the approach may thus not be acceptable for high-reliability applications. Reverse-engineering the bitstream to manipulate the circuit systematically is possible [107], but also seems to be not appropriate for high-reliability applications, as there is no guarantee that all effects have been considered. The only known tool for manipulating synthesized circuits provided by FPGA manufacturers is JBits [43], but it is restricted to the Xilinx Virtex-II series and based on Java. It therefore requires a powerful processor, which excludes it from e.g. space applications, and consequently also methods using this tool, e.g. [36], are not considered to be appropriate for the scope of this thesis.

The embryonics approach [124] addresses self-repair in integrated circuits but requires a particular new kind of FPGA, which is, however, not commercially available.

Another approach for self-repair that uses simple spare gates and is applied on transistor level is presented in [68]. This method provides a deterministic approach towards a reconfiguration, however, it does not tackle the detection of errors.

A method particularly developed for asynchronous circuits is presented in [102]. It is based on fault tolerant basic elements called Pre-Charge Half Buffers, which can be used to construct most asynchronous circuits. The approaches in [68] and [102] are applied on transistor level and thus cannot be implemented in existing programmable devices.

Most of the approaches are not able to transfer/recover the internal circuit state. As a consequence, the application has to be reset and restarted from a ground state. All presented methods can cope with multiple faults with different granularity.

Table 3.1 summarizes the different methods and compares them with respect to some relevant criteria.

Table 3.1: Comparison of Reconfiguration Methods

	sect.	synth.	runt. rec.	state rec.	device	cfg I/F	logic	interc.	algorithm	granul.
DRAFT	§3.2.2	pre	yes	yes	FPGA	FPGA	s, c	partially	deterministic	fine
Fine-Grained	§3.2.3	pre	yes	yes	FPGA	FPGA	c	partially	deterministic	fine
Dynamic Reconf AFTB	§3.2.4	pre	yes	no	FPGA	FPGA	s, c	yes	deterministic	fine
Column-Based Pre-Compiled	§3.2.5	pre	yes	no	FPGA	FPGA	s, c	partially	deterministic	large
Roving STARS	§3.2.6	pre/run	yes	yes	FPGA	FPGA	s, c	yes	deterministic	fine
Node Covering	§3.2.7	pre	no	no	FPGA	FPGA	s, c	yes	deterministic	fine
Shifting Configuration Data	§3.2.8	run	no	no	FPGA	FPGA	s, c	yes	deterministic	medium
Embryonics	§3.2.9	run	yes	yes	FPGA	FPGA	s, c	yes	search	fine
Genetic Algorithms	§3.2.9	run	yes	no	FPGA	FPGA	c	no	search	fine
Reconfigurable Logic Blocks	§3.2.10	run	yes	no	FPGA	prop.	s, c	yes	search	variable
Self-Healing Asynchronous Arrays	§3.2.11	run	yes	no	prop.	prop.	a	yes	search	fine

abbreviations:
sect.: reference to document section with detailed description
synth.: determination of new configuration; pre = pre-defined, run = during runtime
runt. rec.: runtime reconfiguration
state rec.: state recovery
device: target device for this approach; prop. = proprietary device
cfg I/F: configuration interface; prop. = proprietary interface
logic: logic type; s = sequential, c = combinational, a = asynchronous (explicitely)
interc.: interconnections; yes means "covered", no means "not covered", partially means that some interconnections cannot be repaired
algorithm: deterministic = reconfiguration is unambiguous, search = "try and error"
granul.: granularity of reconfiguration; fine = CLB level or below, large = columns/larger areas of FPGA, variable = user-defined

Chapter 4
Self-Healing Approach

4.1 Introduction

The self-healing architecture presented here aims to recover a circuit from permanent faults occurring any time during operation on any position in the circuit. The architecture is based on asynchronous logic and relies on the following properties:

1. **The inherent robustness of asynchronous circuits against transient faults**: This does not mean that transient faults can be neglected, but allows to separate the investigation and evaluation of the fault tolerance against transient and permanent faults into two distinct topics. The fault tolerance with respect to transient faults is handled in [28], whereas the fault tolerance with respect to permanent faults is topic of this thesis.

2. **The fundamental property of asynchronous logic to stop in case of permanent faults**: This is an important advantage of asynchronous circuits, as it implies an inherent fail-stop behavior for permanent faults, which makes fault detection rather simple. Exceptions of this property are highlighted in chapter 2.

3. **The fundamental property of asynchronous logic to autonomously start working as soon as a valid data and acknowledge path exists and consistent data in the pipeline is available**: This implies that we do not need a ground state [73] for re-synchronization, and we can assume that the circuit either is working correctly, or it stops (Property 2). Again, the limitations of this property are presented in chapter 2.

4. **The delay insensitivity of asynchronous logic**: This property is useful with respect to circuit routing and corresponding timing: even if a modified circuit routing results in a different timing, this does not affect the correct function of the circuit. The limitations of delay insensitivity are well known and can be handled very well [79], and are thus not treated any further in this thesis.

The self-healing approach does not seek for a real-time circuit recovery, but to bring the circuit back to operation "as soon as possible" and keep the circuit alive "as long as possible". The idea behind this approach is that e.g. for interplanetary space applications with mission times of several years, the most important issue is to keep the hardware working, while it will be acceptable to wait some seconds or even minutes until the application resumes operation after a fault situation occurs.

The architecture also aims for multiple fault tolerance by adding distributed redundancy with user-defined granularity.

The concept is based on three steps:

1. Try to mitigate the faults by making use of the inherent properties of FSL and some additional measures to reduce transient fault sensitivity

2. If the fault cannot be mitigated, i.e. if it is a permanent fault or the fault leads to a permanent error, force a deadlock

3. Reconfigure the circuit by applying the presented concept

The self-healing approach expects a deadlock, i.e. step 2, as prerequisite, and focuses on step 3 - the circuit reconfiguration.

4.2 Architecture Overview

4.2.1 Concept and Fault Hypothesis

In general, an asynchronous circuit is structured into a pipeline with logic blocks between the register stages. The presented approach adds a redundant pipeline path and extends the asynchronous pipeline by a *reconfiguration unit*, which consists of a deadlock detector (watchdog) and a reconfiguration controller. The watchdog monitors the circuit's activity. After a period of inactivity longer than the watchdog timeout, a deadlock is recognized and a reconfiguration of the circuit is started. Two reconfigurable elements are used: *Self-Healing Cells* (SHC) for the logic function, and *acknowledge switches* for the handshake signals. The reconfiguration controller changes the circuit routing in such a way, that the defective element is isolated and its function is provided by a redundant resource. This concept is shown in Figure 4.1.

As long as data is moving through the pipeline, the phase detectors and acknowledge signals will change their states regularly. This behavior can be used as indication that the circuit is working. Any permanently inconsistent FSL signal vector, e.g. caused by a permanent error, stops the handshake and thus leads to a *deadlock*. In this context we consider permanent faults using the stuck-at fault model (see section 2.1.3). The class of permanent faults represents all faults with a permanent manifestation. This could be hardware defects or e.g. faults caused by radiation, such as total dose effects, which are a common problem in space missions [24, 9].

During a deadlock the input values to the register or combinational logic, which is affected by a permanent fault, are kept valid, because no subsequent pipeline stage will consume the data due to the stopped handshake (see section 2.4). In general, as soon as a new configuration is found that establishes a valid data and acknowledge path, the circuit will continue its operation autonomously without loss or corruption of data. However, one case has been identified where this assumption is not true (see section 2.4.5). This exception is a general problem of FSL and is not related to the self-healing concept itself. However, because no deadlock occurs in this case, such an effect could generate erroneous results which cannot be detected by only applying the presented concept.

The circuit is defined to be working if at least one of the two redundant outputs shows a correct result. The determination of the correct output, as well as the treatment of erroneous results, need to be covered on application level and are not scope of this thesis.

4.2. Architecture Overview

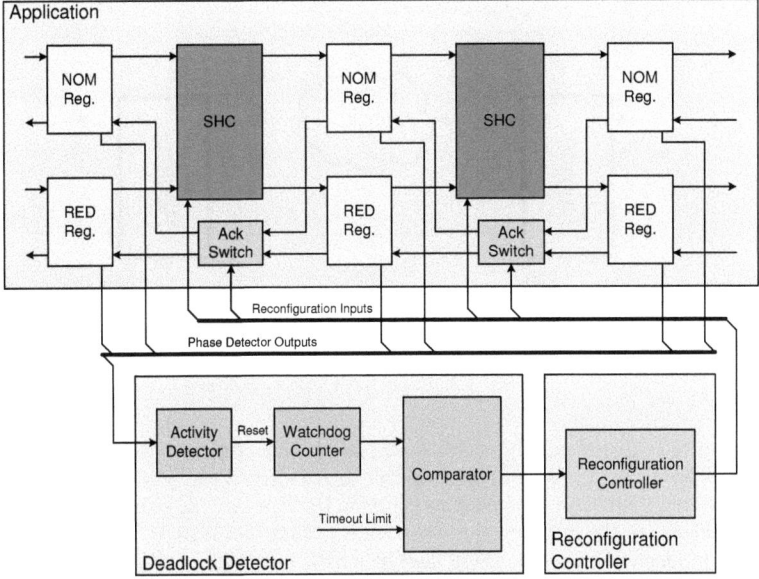

Figure 4.1: Self-Healing Approach

4.2.2 Fault Locations

Figure 4.2 shows – at the example of a standard FSL pipeline – that all possible fault locations can be subsumed by three fault positions: (i) the input to the combinational logic, (ii) the input of registers, and (iii) the acknowledge signal. All faults within the blocks (combinational logic, register/latch control and faults within the register/latch, [103]) will end up with one of these three effects (indicated with dotted lines).

The same is true for a self-healing redundant pipeline. As the self-healing architecture expects a deadlock as prerequisite for a reconfiguration, i.e. the fault has settled and shows effect, all fault conditions within the pipeline and within the involved elements can be subsumed by the following three cases:

1. Faults at the input of a SHC (covering those faults within the SHC that affect both outputs, faults at the register outputs, and the interconnection between registers and SHCs)

2. Faults at the input of a register (covering faults within the SHC affecting one output, single faults at the SHC outputs and on the interconnections between SHCs and registers)

3. Faults at an acknowledge signal (covering faults on the acknowledge output/input at the register component, interconnections between the registers and faults in the acknowledge switches - if applicable)

4. Self-Healing Approach

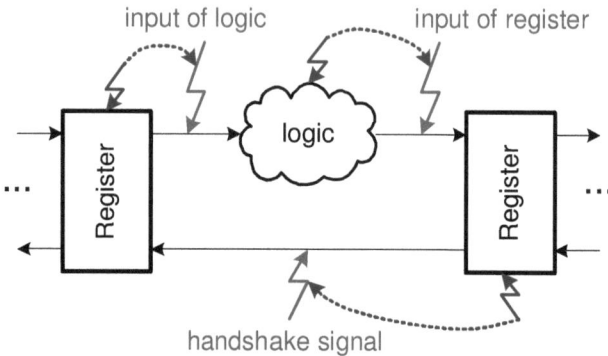

Figure 4.2: Expected Fault Locations

Faults within registers or SHCs will also end up with one of the described effects. Depending on the fault position, the configurable elements must be controlled in different ways.

Faults within a SHC can cause various effects in the pipeline, and either one or both outputs can be incorrect. In the first case this situation is similar to a fault at a register input, in the latter case it affects the pipeline like a fault at a SHC input. The correct identification of the fault has to be taken into account for the design of the reconfiguration unit (see section 4.4.5 for observable symptoms).

Faults within registers will always affect either the data output (equal to a fault at the input of the subsequent logic) or the handshake signal to the preceding register.

The following paragraphs describe the means to handle the different faults in the self-healing architecture.

4.2.3 Reconfiguration of Combinational Logic

To be able to re-route the circuit so that the defective resources are bypassed, switching possibilities are needed. For the logic circuits *Self-Healing Cells* (SHCs) were introduced (Figure 4.3a).

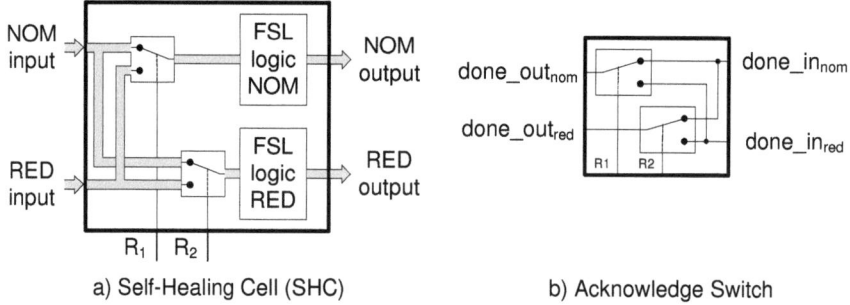

Figure 4.3: Self-Healing Cell (SHC) and Acknowledge Switch

44

4.2. Architecture Overview

A SHC is an internally redundant circuit element, which provides flexible internal routing possibilities while maintaining the same external interface. Depending on the configuration, the redundant logic circuits implemented in the SHC either use the nominal or redundant inputs and provide two independent outputs. The routing is controlled by reconfiguration inputs. One SHC is able to tolerate at least one fault, either internally or at its interfaces. Depending on the distribution of the faults over the whole application there is a significant probability that even multiple faults can be mitigated (section 5.3.1, [97]). The logic circuit in the SHC can be of arbitrary complexity, ranging from low level gates (AND, OR, etc.) as shown in Figure 4.4a, up to complex circuits (e.g. arithmetic units) such as in Figure 4.4c. Individual SHCs can be composed to construct larger and more complex circuits (Figure 4.4b), which permits establishing a hierarchical design. This allows to implement a self-healing logic at different granularity, which influences the reconfiguration possibilities and fault tolerance.

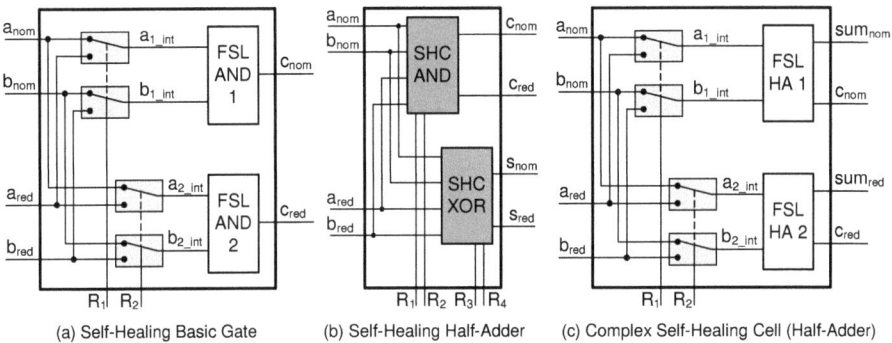

Figure 4.4: Coarse and Fine Granular SHCs

Apart from the granularity of the logic function per block, also the reconfiguration inside the block can be performed in different granularity, depending on the number of switching possibilities between the nominal and redundant signals within a SHC. The results of the analysis can be found in section 5.3.2. A feasible trade-off is to use two switches, which allows to select whether the nominal or redundant signals shall be used as source for the nominal and redundant logic. All further explanations in this chapter assume this SHC architecture.

4.2.4 Reconfiguration of Control Logic

Just by applying SHCs, faults of combinational functions can be recovered. However, in an asynchronous pipeline, a permanent fault at a register input may block the handshake. So, additional elements called *Acknowledge Switches* are needed (Figure 4.3b) [98], to be able to re-route the acknowledge signal around a defective register as shown in the example in Figure 4.5.

4. Self-Healing Approach

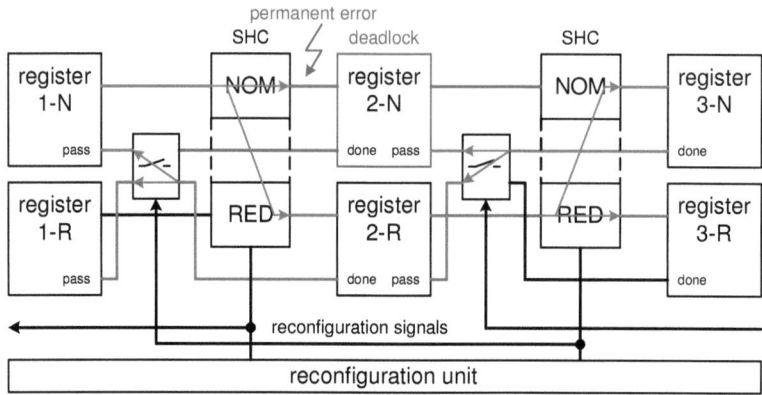

Figure 4.5: Handshake Reconfiguration

4.2.5 Reconfiguration Unit

The reconfiguration process is controlled by the *reconfiguration unit*, which consists of a *deadlock detector* and the *reconfiguration controller*.

Deadlock Detector

To detect a deadlock, some kind of watchdog is used that will be reset by the circuit's activity, e.g. by one of the handshake signals. Unfortunately, a deadlock is not distinguishable from the state where the circuit waits for transitions to complete or for new data to arrive at the pipeline input. Therefore the unbounded delay model in QDI circuits (see chapter 2) has to be violated. If there is no transition within the watchdog's timeout, it can be assumed that a deadlock has occurred. Strictly speaking, the watchdog timeout restricts the maximum allowed delay in the system to some upper boundary, which is in contradiction to the unbounded delay model. However, as long as this timeout is several orders of magnitude higher than the actual maximum system delay, the unbounded delay model still holds at least within practical limits. Choosing a large timeout basically is considered acceptable from a fault-tolerance perspective, since the faults are assumed not to occur frequently. For example, in an integrated circuit the delay lies within the pico or nanosecond range. If a watchdog is designed that requires a microsecond or even a millisecond to expire, the circuit delays won't even come close to the watchdog timeout – except there is something wrong. Such a timeout is still sufficient to provide the circuit recovery within reasonable time.

Reconfiguration Controller

The task of the reconfiguration controller is to reconfigure the circuit by applying appropriate patterns to the reconfiguration inputs of the SHCs and acknowledge switches, so that the defective resource is bypassed. In order to have an asynchronous circuit operating, both the data *and* the handshake path must be valid. Depending on the fault location, it is necessary

4.2. Architecture Overview

to either reconfigure a SHC, or the handshake signal(s), or both. The required activities are explained in detail in section 4.3.

Each time the watchdog expires, the reconfiguration controller is triggered and changes the circuit routing. If the new configuration does not solve the problem or another fault occurs, the watchdog expires again. This sequence is repeated until a working solution is found. Basically, using only the acknowledge signals to detect a deadlock does not provide any information about the fault location. If the phase detector outputs of the adjacent stages are used, the defective pipeline stage can be determined and the fault location can even be narrowed to the affected element (see section 4.4). This information can be used to define an algorithm that reconfigures the circuit in a directed manner until it resumes its operation. Results of the experiments using different reconfiguration controllers can be found in section 5.4.2.

Note that in the figures usually a single SHC between two registers is drawn, but this is only for illustration. A fine granular logic implementation would consist of several SHCs so that knowing the defective pipeline stage does not give more detailed information about the fault location within this SHC.

Tradeoffs

For the actual implementation of the reconfiguration unit several aspects have to be considered. The following explanations are not quantified in this place, but summarize thoughts on the self-healing design. Some issues are treated in other sections and the relevant references are given below, others are left open for future work.

The system immanent application processing time, i.e. the time from a new data input until there occurs activity on the handshake signals, determines the lower boundary of the watchdog timeout and thus directly influences the reconfiguration time. Shorter timeouts would trigger the reconfiguration unit even in a fault-free case, and as the pipeline is in a dynamic state, unexpected reconfigurations during pipeline operation could occur based on the observed signals.

The fault tolerance requirements, e.g. fault model (fault types, total number of faults and fault frequency, see section 2.1), set up the requirements for the "intelligence" of the reconfiguration controller, and define the needed granularity of the SHC internal logic as well as the architecture of the reconfiguration unit (distributed or central).

A complex reconfiguration controller, that evaluates a lot of details of the pipeline internal signals, might be able to handle more faults and fault combinations, but at the same time it will require more resources to handle the increased state space and thus it might be getting more susceptible to faults than a simple controller that uses fewer resources.

A fine granular circuit implementation will provide higher fault-tolerance (in particular against multiple faults), but requires (much) more reconfiguration signals that increase the resource effort for the reconfiguration controller (see section 4.6).

Basically, it would be sufficient to have one central reconfiguration unit which handles all reconfiguration inputs. However, if the complexity is high and/or the circuit is very large, it might be easier to use a distributed, modular reconfiguration unit. Each unit will then cover only a small part of the circuit. This approach allows to define self-contained regions in the design that can be reconfigured independently. Furthermore, the regions can be built with different approaches (e.g. fine/coarse granular) according to their criticality in the system (e.g. reconfiguration time).

4.2.6 Self-Healing Reconfiguration Unit

Faults occurring within the reconfiguration unit have not been addressed so far. Although this is not the main scope of the thesis, some words are spent on this issue.

Basically, since the reconfiguration unit can be designed in asynchronous logic as well, the same self-healing concept as for the application logic can be applied. At least two reconfiguration units would be needed as shown in Figure 4.6a: RU-A covers the application and the support unit RU-B, and RU-B covers the main unit RU-A. From the view of RU-A, RU-B acts like an extension of the application pipeline. All circuits (application, RU-A and RU-B) are operating independently from each other. Any fault in RU-A, which leads to a deadlock in this unit, will be detected and handled by RU-B without interruption of the application. A fault in RU-B, however, will not be handled unless also a fault in the application occurs. Then both the application and RU-B will be reconfigured by RU-A.

To overcome the disadvantage that the application needs to fail in order to repair a fault in the reconfiguration unit, an architecture as in Figure 4.6b could be used. Here RU-A checks the application, RU-B checks RU-A and RU-C, and RU-C checks RU-B. The application can thus operate independently from the reconfiguration units and vice versa. However, a fault only in RU-A would not lead to a reconfiguration unless also RU-C stops its operation. If the application fails before RU-C, no repair would be started and the whole application would fail.

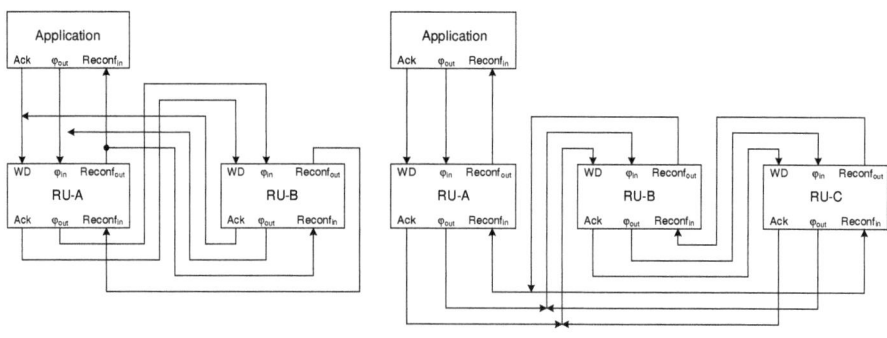

a) Self-Healing Reconfiguration Unit, Variant 1   b) Self-Healing Reconfiguration Unit, Variant 2

Figure 4.6: Self-Healing Reconfiguration Unit

With these approaches the application and the reconfiguration unit are not independent from each other. Since the reconfiguration unit does not contribute to the intended function of the application, the whole circuit is not free of single-point failures from an application point of view.

Basically, it is a general problem of fault-tolerant systems that implemented redundancy does not help if the checker and/or controller fails. However, while e.g. in a TMR system a failing voter already results in a system failure, the described approaches - although they look awkward at a first glance - would autonomously bring the system back to operation.

The effort of building the reconfiguration unit in a self-healing way can be avoided if it provides significantly higher reliability than the application. As the application in general will dominate the complexity and resource occupation, this goal should easily be achieved. The same conclusion has been reported for other approaches, e.g. in [70].

To avoid common mode failures, the application and the fault-tolerance logic can be distributed into different components, e.g. as described in [121].

4.2.7 Default State of Configurable Elements

The combination of SHCs and acknowledge switches allows various configurations of a pipeline, where data and handshake path could be mixed arbitrarily. Although this basically gives a lot of possibilities to bypass a defective element, the crossing of these paths can cause severe troubles due to timing differences. As long as the handshake path follows the associated data path, all timings (time of capturing data in a register, time of asserting the acknowledge, etc.) are defined from the same source and from the same path. Switching either the data or the handshake path to the redundant pipeline path means that the nominal data timing is defined by the redundant handshake timing or vice versa. In this configuration timing assumptions are required for correct functionality – otherwise the reconfiguration might fail (see section 5.5).

By default, a configuration for the SHCs was chosen where both outputs are determined from the nominal input. In order to be able to use the same circuit for reconfiguration with and without acknowledge switches, the default configuration for the acknowledge switches was chosen to have individual acknowledge paths for the nominal and redundant pipeline. This was appropriate for the reconfigurations without acknowledge switches, but caused timing problems with acknowledge switches, so that also a different default configuration was used for the experiments (see section 5.5).

Details about the timing in different pipeline configurations are discussed in section 4.3.5.

4.2.8 Transformation from FSL to SH-FSL

The steps to transform a standard FSL circuit into a Self-Healing FSL circuit are as follows:

- Duplicate the pipeline into a nominal and a redundant pipeline
- Replace each combinational logic block by a SHC circuit of appropriate granularity
- Add an Acknowledge Switch for each handshake connection between two registers[1]
- Add a reconfiguration unit to control the reconfiguration inputs

These steps are depicted in Figure 4.7. Note that the first step also implies the duplication of the registers.

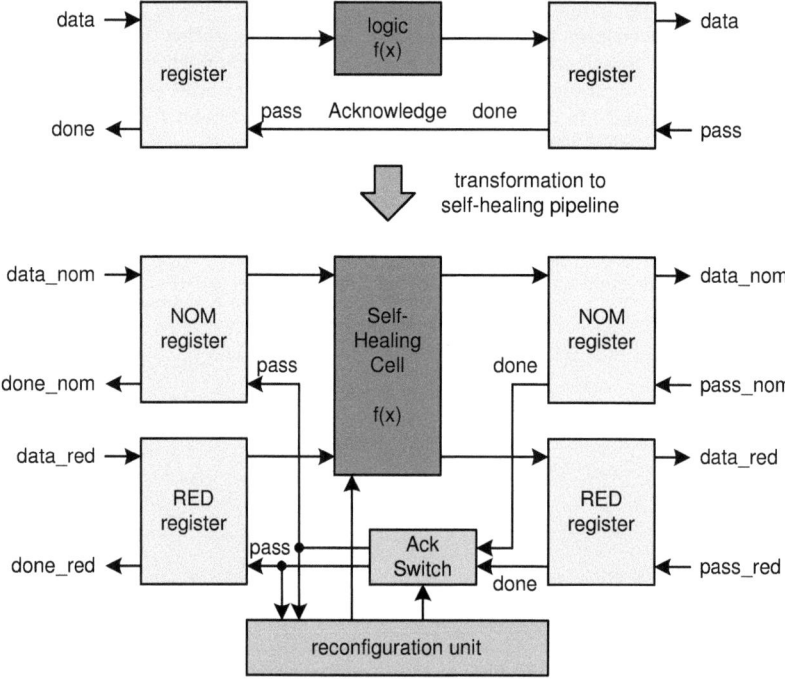

Figure 4.7: Transformation from FSL to SH-FSL Circuit

---

[1]Note: As will be shown later, acknowledge switches are not mandatory, but increase the reconfiguration possibilities and thus the fault tolerance.

4.3 The Principle of Pipeline Reconfiguration

4.3.1 Introduction

The self-healing architecture shall be able to repair stuck-at faults on any position in the circuit while losing as few resources as possible in order to keep the maximum of possibilities for further repairs. In section 4.2 it was shown that all faults can be subsumed by three fault locations.

The following paragraphs describe each of these cases in detail and present possible configurations to recover from the particular fault situations. For these explanations a single fault is assumed, corresponding to the first fault in the pipeline. The possibilities for successful reconfigurations of subsequent faults depend on the reconfiguration algorithm, i.e. how the first fault has been handled. This topic is treated in section 4.4.6.

Note: the figures referred to in the following paragraphs are presented at the end of this chapter.

The following default settings are assumed in the explanations:

- SHC: the nominal input is used as source for both the nominal and redundant logic. Consequently, the faults are injected in the nominal path. Any (first) fault in a redundant path will not influence the result on the nominal path in this configuration.

- Acknowledge switch: each pipeline uses its associated acknowledge, i.e. the nominal path receives the nominal acknowledge and the redundant path the redundant acknowledge. This configuration is the same as for a pipeline without using acknowledge switches which eases the comparison. As the explanations only address the principle possibilities for routing both the data and acknowledge path, regardless of their usability in a real circuit, the default configuration does not matter in this context. As will be described later in section 5.4.3, it turned out to be better if the acknowledge path follows the data path for reconfigurations using acknowledge switches. This is also consistent with the standard way of designing asynchronous pipelines.

Notice that reconfigured components (differing from the default configuration) are shaded in grey. The fault origin is designated with a circled cross "$\otimes$", blocking signals being a consequence of the initial fault are designated with a bold cross "$\times$", and other non-nominal signal states due to deadlock propagation by a thin cross.

Furthermore, the pipeline status in the fault conditions will be explained. Knowing this status is a prerequisite for defining rules for a reconfiguration unit (see section 4.4).

4.3.2 Faults at SHC Inputs

FSL logic will only update its output if consistent data is applied to the inputs (see section 2.4.2). Assuming a SHC with two reconfiguration inputs (see also section 5.3.2), four SHC configurations are possible. The following explanations consider a fault at the nominal input of a SHC. The same arguments, however, are also true for faults at the redundant input for the inverse configuration.

1. Configuration 1 (default configuration): Both SHC outputs use the nominal input as source (Figure 4.18a). A fault at the nominal SHC input will cause both outputs to keep the old value, i.e. the pipeline will end up in a deadlock. A fault at the redundant input will not influence the outputs in this configuration.

4. Self-Healing Approach

2. Configuration 2: Both SHC outputs use the redundant input as source (Figure 4.18b). A fault at the nominal input does not influence the outputs. Both the nominal and redundant pipeline path keep working and all handshakes can be maintained. A fault at the redundant SHC input will cause both outputs to keep the old value, i.e. the pipeline will end up in a deadlock.

3. Configuration 3: The nominal output is defined by the nominal input, the redundant output by the redundant input (Figure 4.18c). Assuming a fault at the nominal input, this (re-)configuration would only recover the redundant path. The nominal pipeline path would be stopped forever, because the nominal handshake is blocked. This can be improved by the acknowledge switch (Figure 4.18d): If the redundant handshake is fed to the preceding nominal register, the nominal pipeline will work up to the faulty SHC. By appropriately configuring the subsequent SHC it is even possible to establish a circuit where both pipeline paths are working again (Figure 4.18e). The fault is locally bypassed and all preceding and subsequent elements are in default configuration. However, this configuration splits the data and acknowledge path, which will only work in highly symmetric circuits (see section 4.3.5). As will be shown later in section 5.5, the forks emerging in such configurations cause troubles in hardware implementations.

4. Configuration 4: In this case the nominal output uses the redundant input and the redundant output the nominal input as source (Figure 4.18f). A fault at the nominal input would cause the pipeline to process one more token on the nominal path, because the redundant output shows valid data, and then end up in a deadlock because the redundant register would receive inconsistent input data and thus not update the handshake signal. This further causes the preceding redundant register to not feed-through the new input data, which means that there will not appear any new data at the input of the nominal register. Therefore the nominal handshake signal must be routed to the redundant preceding register by appropriately configuring the acknowledge switch (Figure 4.18g). However, as explained in section 4.3.5, crossing the data and acknowledge path can cause troubles and such a configuration is not recommended. Without using acknowledge switches the fault at the SHC input cannot be repaired in this SHC configuration.

The figures show abstract SHCs where the two paths, nominal and redundant, can be switched as a whole. For a fine granular implementation, where the combinational circuit is built from several SHCs, this is basically also possible, but several switches of the involved SHCs would have to be switched to achieve such a configuration within the pipeline. The four described fault cases still remain true also for fine granular circuits, since also internal faults (within the SHC) will finally have the same effect at the circuits outputs (e.g. that one of the outputs will not update its value). For a fine granular circuit it might even be possible to find a working configuration without switching the whole path. This, however, depends on the granularity and the design of the circuit. A comparison of fine and coarse granular SHC reconfiguration and the resulting fault tolerance is presented in section 5.3.1.

4.3. The Principle of Pipeline Reconfiguration

4.3.3 Faults at Register Inputs

A fault at a register input causes the register to stop updating the output with new input data. In the default configuration of the SHCs and acknowledge switches this would block both the nominal and redundant path (Figure 4.19a). While for faults at SHC inputs there exist solutions where only the SHC has to be reconfigured for a local repair, the situation is more complex for faults at register inputs. Without acknowledge switches the affected pipeline path is completely lost with the simple reconfiguration shown in Figure 4.19b. Any additional fault in the other pipeline path cannot be repaired any more. In the optimized solution the pipeline path is lost up to the defective register (Figure 4.19c). Here, all SHCs up to the fault location are configured to separate the nominal and the redundant path, and after the fault location the default configuration is used again. With acknowledge switches in addition to SHC reconfigurations the fault can be locally bypassed and all preceding and subsequent elements are in default configuration (Figure 4.19d). Such a configuration can, however, be troublesome (see section 4.3.5).

Note that with the configuration in Figure 4.19b the nominal and redundant pipeline path are working completely independent from each other. Since there will be slight timing differences, the nominal and redundant pipeline outputs would drift apart after some time and experience a mis-alignment. Thus, at least one data source in the path has to generate synchronized outputs. This can be a SHC configured to use the same source for the nominal and redundant logic, or a common pipeline input.

4.3.4 Faults at Acknowledge Signals

Faults at acknowledge signals cause quite similar effects as faults at register inputs. There exist two basic solutions: (i) using only SHCs for reconfiguration, the affected pipeline path is lost (Figure 4.20b), or (ii) acknowledge switches are used, which allows a local repair without loss of a huge amount of resources. In this case the acknowledge signal of the redundant register is fed back to the nominal register and all components and both paths stay in operation (Figure 4.20c). The same optimization as for register faults without using acknowledge switches can be applied and is shown in Figure 4.20d. As described in section 4.3.5 below, splitting the data and acknowledge path can cause timing problems.

4.3.5 Timing Investigation of Different Pipeline Configurations

This section exploits the influence of different pipeline configurations on the pipeline timing. In section 2.4.4 two basic constraints were defined in order to guarantee a correct capturing of data in registers. In a standard non-redundant pipeline these constraints are easy to fulfill because data and acknowledge signals are provided by the same element (register). For a self-healing pipeline the situation is a bit more complex, as - depending on the configuration - data and acknowledge signals are not necessarily defined by the preceding or subsequent register of the same path, and thus the constraints could be violated.

In the redundant pipeline each of the timing parameters defined in section 2.4.4 can be expressed for both the nominal and the redundant path (e.g. t_{R_Nom}, t_{R_Red}).

Three particular configurations are taken as examples and described in detail below:

4. Self-Healing Approach

- Pipeline with acknowledge switches in nominal configuration, i.e. both the data and handshake path are defined by the nominal pipeline. This is the default configuration for the pipeline using acknowledge switches.
- Pipeline without acknowledge switches in nominal configuration, i.e. the data path is defined by the nominal pipeline, but both pipelines maintain their individual handshake. This is the default configuration for the pipeline without acknowledge switches.
- Pipeline without acknowledge switches and one SHC configured to use the redundant input as source for both the nominal and redundant output. This configuration caused troubles during the hardware experiments (refer to section 5.5).

The three listed configurations cover all situations being used with the reconfiguration algorithms described in this thesis. Several other configurations are possible (e.g. one or more SHCs within the pipeline splitting the nominal and redundant path) which could make the timing conditions even more severe, as e.g. timing differences could accumulate from multiple stages. These situations need to be investigated in detail if they shall be used with a more complex reconfiguration algorithm, however, this is beyond the scope of this thesis.

In the following explanations the timing is investigated from the view of an input to a SHC which is used as source for both the nominal and redundant output and thus acts as synchronization point in the pipeline. For the registers that receive the data and acknowledge from registers of the same path, the same equations as listed in section 2.4.4 for the standard pipeline apply. The equations given in the following paragraphs describe the situation for registers, where data and acknowledge are provided by different elements.

The capturing of a new token applied to the input of the SHC is finished after the token has propagated through the SHC and the register and the register has assigned the *Done* signal. This time is designated as $t_{capture}$.

Equation 5 $t_{capture}(n)(nom) = t_{L\_Nom}(n) + t_{R\_Nom}(n) + t_{H\_Nom}(n)$

Equation 6 $t_{capture}(n)(red) = t_{L\_Red}(n) + t_{R\_Red}(n) + t_{H\_Red}(n)$

Pipeline with AS - Synchronized Data and Acknowledge Path

In the pipeline configuration shown in Figure 4.8 the timing of both the data and the acknowledge path is defined by the nominal pipeline, as the nominal register provides the data to the SHC, the SHC uses the nominal data input only and the nominal *Done* signal is fed back to the *Pass* input of both the nominal and the redundant register.

According to Constraint 1, register n must not receive an acknowledge from register $(n+1)$ before it has received and acknowledged the last token. The time $t_{capture}$ for the redundant register n must therefore be shorter than the time the token needs to propagate to the output of the nominal register $(n+1)$ and the register assigns the *Done* signal.

Constraint 3 $t_{capture}(n)(red) < t_{L\_Nom}(n) + t_{R\_Nom}(n) + t_{capture}(n+1)(nom)$

Constraint 2 requires that register n must not receive a new token at its input before the last token has been captured and acknowledged. The redundant register n thus must have finished its capturing of token i before the nominal register n has requested a new token $(i+1)$ from register $(n-1)$ and this token has propagated to the input of the redundant register.

Constraint 4 $t_{capture}(n)(red) < t_{capture}(n)(nom) + t_{R\_Nom}(n-1) + t_{L\_Red}(n)$

4.3. The Principle of Pipeline Reconfiguration

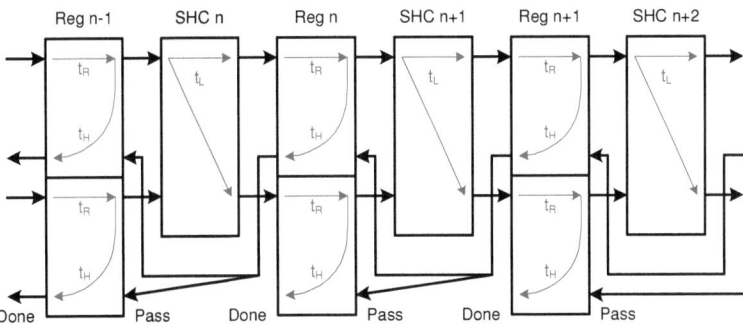

Figure 4.8: Pipeline with Acknowledge Switches and Synchronized Data and Handshake Paths

Pipeline without AS - Split Data and Acknowledge Path

In the pipeline configuration shown in Figure 4.9 the timing of the data is defined by the nominal pipeline, as the nominal register provides data to the SHC and the SHC uses the nominal input only, while the acknowledge path is defined individually by the nominal and redundant registers.

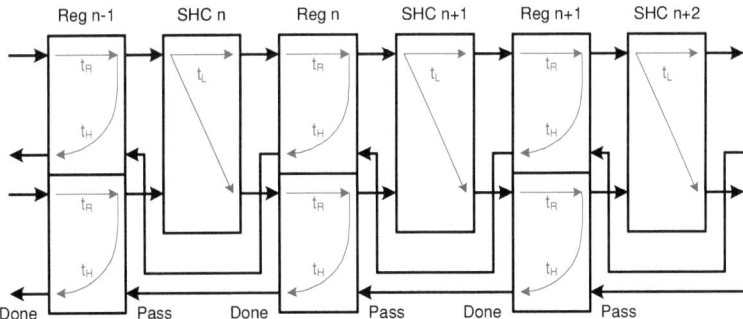

Figure 4.9: Pipeline without Acknowledge Switches and Split Data and Handshake Paths

To fulfill Constraint 1, the redundant register n must have finished its capturing before the nominal register $(n+1)$ provides the *Done* signal.

Constraint 5 $t_{capture}(n)(red) < t_{L\_Nom}(n) + t_{R\_Nom}(n) + t_{capture}(n+1)(red)$

To fulfill Constraint 2, the same logic applies as for the pipeline with acknowledge switches.

Constraint 6 $t_{capture}(n)(red) < t_{capture}(n)(nom) + t_{R\_Nom}(n-1) + t_{L\_Red}(n)$

Pipeline without AS - Problem Case

The pipeline configuration shown in Figure 4.10 is basically the same as before, but with one SHC configured to use the redundant input as source for the nominal and redundant output. Both pipeline paths maintain their individual handshake signals. This configuration may be the result of a self-repair (see section 5.5).

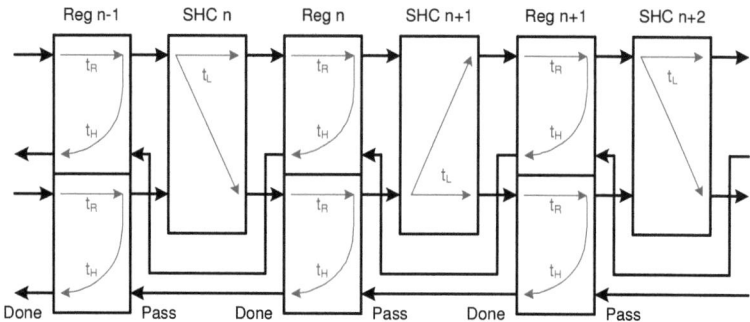

Figure 4.10: Pipeline without Acknowledge Switches, Problem Case

The redundant register n provides the data to SHC $(n + 1)$ and consequently to register $(n + 1)$. In order to fulfill Constraint 1 the nominal register n therefore must have finished the capturing before the token has propagated through the redundant register n and SHC $(n + 1)$ to the nominal register $(n + 1)$ and this register has assigned its *Done* signal.

Constraint 7 $t_{capture}(n)(nom) < t_{L\_Red}(n) + t_{R\_Red}(n) + t_{capture}(n+1)(nom)$

Regarding Constraint 2, two different constraints can be defined. From the view of SHC n the redundant register n must have finished its capturing of token i before the nominal register n requests a new token $(i + 1)$ and this token has propagated to the input of the redundant register n.

Constraint 8 $t_{capture}(n)(red) < t_{capture}(n)(nom) + t_{R\_Nom}(n-1) + t_{L\_Red}(n)$

From the view of SHC $(n+1)$ the nominal register $(n+1)$ must have finished its capturing of token i before the redundant register $(n+1)$ requests a new token $(i+1)$ and this token has propagated to the input of the redundant register n.

Constraint 9 $t_{capture}(n+1)(nom) < t_{capture}(n+1)(red) + t_{R\_Red}(n) + t_{L\_Nom}(n+1)$

4.3. The Principle of Pipeline Reconfiguration

Comparison and Discussion of Timing Investigation

In all configurations particular timing constraints for the provision of new data and the acknowledge signal must be considered to ensure a correctly capturing of data in all registers. All these conditions essentially bound the skew between the $t_{capture}$ of adjacent stages to the sum of some t_R and t_L. The delay t_L through a SHC is usually significant, so the conditions are relatively safe in practice.

While in a non-redundant pipeline the conditions can easily be achieved (refer to section 2.4.4), in the self-healing pipeline the mixture of signal sources between the nominal and the redundant pipeline path requires more complex constraints. The timing conditions emerge from the existence of concurrent paths. In the fault free configurations these paths are short and span two pipeline stages only (this was one important reason for this choice of initial configuration), but after repair longer concurrent paths may form, and hence the potential for violations of the timing constraints increases.

With the set of equations expressed above it can be shown that for arbitrary timings there exist several combinations that will lead to a violation of one or more constraints and consequently to an unexpected pipeline behavior as some tokens might not be captured. Basically even meta-stabilities could occur if the inputs are changing while the register is in a dynamic state. However, such a behavior has not been observed.

4.3.6 Summary

Basically, the first fault on all three fault locations can be repaired with and without the use of acknowledge switches. However, the amount of resources that are lost differs significantly. This influences the fault tolerance with respect to multiple faults, since with more operational resources there remain also more possibilities for subsequent reconfigurations.

Without acknowledge switches fault combinations affecting the nominal and redundant path at the same time can only be repaired if the faults are particularly distributed. Otherwise such a situation leads to the loss of the whole pipeline. With acknowledge switches only those combinations affecting nominal and redundant elements at the same position within the pipeline (e.g. nominal and redundant part of the same register) or neighbor elements (e.g. nominal SHC input and redundant acknowledge of preceding register) lead to a loss of the circuit. Details about these cases are presented in section 4.5.

Care has to be taken that there is at least one synchronization point between the nominal and the redundant path in the pipeline, otherwise a mis-alignment of the pipeline outputs can occur due to the accumulation of the individual pipeline timings.

As will be shown in section 5.4.2, only a subset of the presented configurations will actually be used for a practical implementation. On the one hand, not all configurations are necessary for the developed algorithms, on the other hand the timing investigation in section 4.3.5 showed that some configurations are difficult to implement with respect to timing constraints.

4.4 Fault Diagnosis in a Pipeline

This section explains how the different fault types described in the previous sections can be identified in the pipeline after a deadlock has been detected. Initially the descriptions assume a nominal (fault-free) behavior of the pipeline prior to the fault occurrence. The validity of this assumption and exceptions for multiple faults are discussed at the end of this chapter.

4.4.1 General

The designators as per Table 4.1 will be used for the explanations. For each of the fault descriptions a table showing the pipeline status will be given. In these tables the "X" designates the fault location. Note that in the columns "φ_{in}" and "φ_{out}" the phase status φ_n equals the output of the phase detectors checking the FSL data vector, while in the columns "$Done$" and "$Pass$" it represents the logical signal value

Designator	Meaning
Reg^m	m^{th} register in the pipeline without further designators both NOM *and* RED registers are meant
SHC^m	m^{th} SHC in the pipeline. The first element in the pipeline is SHC^1. SHC^2 follows Reg^1.
Ack^{mk}	Acknowledge signal between register Reg^m and Reg^k
$Element_{in|out}$	input|output of $Element$
$\overline{Signal}$	faulty $Signal$ e.g. $\overline{Ack^{mk}}$, $\overline{SHC_{in}}$
φ_n	n^{th} token with phase φ; $\varphi_n \neq \varphi_{n+1}, \varphi_n = \varphi_{n+2}$

Table 4.1: Designators

A generic pipeline with the numbering scheme as in Figure 4.11 is analyzed. The pipeline receives data from two independent source buffers and stores the data in two independent sink buffers. The sink buffers mirror the phases of their individual data as *Done* signal to the last register in the pipeline. All explanations assume the pipeline and its elements, respectively, to be in their default configuration as described in section 4.2.7.

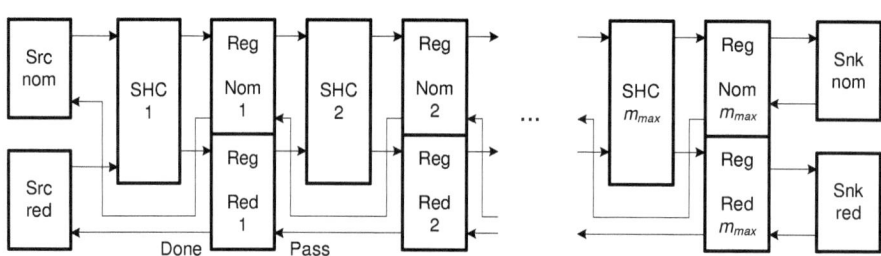

Figure 4.11: Generic Pipeline for Symptom Determination

4.4. Fault Diagnosis in a Pipeline

4.4.2 Fault at SHC Input

Faults at SHC inputs will generate inconsistent input data for the FSL combinational logic. In the assumed strongly indicating implementation, FSL logic will only change its output if consistent data is applied. Thus, the SHC will keep its output corresponding to the last consistent input. Since the SHC is embedded in a pipeline, the subsequent register will not receive new data with a new phase and stop the handshake.

In this case the input and the output phase detectors of register Reg^{m+1}, located subsequently to the faulty SHC^{m+1}, will indicate the same phase, since still the old input data is applied and the register has already captured these data.

The register Reg^m, preceding the faulty SHC, will capture the new input data with phase φ_{n+1}. Then the pipeline enters the deadlock status, as the new data will not be processed by SHC^{m+1} and thus Reg^{m+1} will not receive and acknowledge the new data. Table 4.2 shows the pipeline status for the deadlock condition due to a fault at the input of SHC^{m+1}. This situation is depicted in Figure 4.18a.

	Reg^m				SHC^{m+1}		Reg^{m+1}	
	φ_{in}	φ_{out}	$Done$	$Pass$	φ_{in}	φ_{out}	φ_{in}	$Done$
NOM	φ_{n+2}	φ_{n+1}	φ_{n+1}	φ_n	X (φ_n)	φ_n	φ_n	φ_n
RED	φ_{n+2}	φ_{n+1}	φ_{n+1}	φ_n	φ_n	φ_n	φ_n	φ_n

Table 4.2: Pipeline Deadlock State for Fault at SHC Input

Since by default the SHC uses the same input as source for both the nominal and redundant logic, and thus both outputs will provide the same result, the same symptoms will occur at both the nominal and redundant register in the self-healing pipeline.

4.4.3 Fault at Register Input

A fault at a register input will cause inconsistent input data, thus the input phase detector will not update its output to a new phase. All preceding registers will capture new input data up to the faulty register, because this one does not update its $Done$ signal. All subsequent registers will have captured the last consistent data. The status as shown in Table 4.3 will appear. The SHCs between the registers will only pass through the nominal data to the nominal and redundant register and are not shown in the table for the sake of brevity.

	Reg^{m-1}				Reg^m				Reg^{m+1}	
	φ_{in}	φ_{out}	$Done$	$Pass$	φ_{in}	φ_{out}	$Done$	$Pass$	φ_{in}	$Done$
NOM	φ_{n+2}	φ_{n+1}	φ_{n+1}	φ_n	X (φ_n)	φ_n	φ_n	φ_n	φ_n	φ_n
RED	φ_{n+2}	φ_{n+2}	φ_{n+2}	φ_{n+1}	φ_{n+1}	φ_{n+1}	φ_{n+1}	φ_n	φ_n	φ_n

Table 4.3: Pipeline Deadlock State for Fault at Register Input

Even with a faulty input at the nominal register Reg_{NOM}^m, the redundant register Reg_{RED}^m can capture the new data provided from SHC^m. However, the subsequent SHC^{m+1} will use the (not updated) nominal output $Reg_{out\_NOM}^m$ as source, so that register Reg^{m+1} will not receive new data and still acknowledges the last consistent data. Consequently, $Done_{RED}^m$ will be unequal to $Pass_{RED}^m$.

4. Self-Healing Approach

A faulty input at a redundant register would block the redundant handshake but not lead to a deadlock in the default configuration of the pipeline, i.e. remain as a latent fault. This issue is described in section 4.4.6.

4.4.4 Fault at Acknowledge Signal (Pass Input)

A faulty acknowledge signal ($Pass$ signal) will block the handshake of the register receiving the faulty signal, which will thus not capture the new input data. All preceding registers will capture new input data up to the affected register, because this one does not update its $Done$ signal. All subsequent registers will have captured the last consistent data. The status as shown in Table 4.4 will appear. The SHCs between the registers will only pass-through the nominal data to the nominal and redundant register and are not shown in the table.

	Reg^{m-1}				Reg^m				Reg^{m+1}	
	φ_{in}	φ_{out}	$Done$	$Pass$	φ_{in}	φ_{out}	$Done$	$Pass$	φ_{in}	$Done$
NOM	φ_{n+2}	φ_{n+1}	φ_{n+1}	φ_n	φ_{n+1}	φ_n	φ_n	X ($\neg\varphi_n$)	φ_n	φ_n
RED	φ_{n+2}	φ_{n+2}	φ_{n+2}	φ_{n+1}	φ_{n+1}	φ_{n+1}	φ_{n+1}	φ_n	φ_n	φ_n

Table 4.4: Pipeline Deadlock State for Faulty Pass Signal

Even with a faulty $Pass$ signal at the nominal register Reg^m_{NOM}, the redundant register Reg^m_{RED} can capture the new data. However, the subsequent SHC^{m+1} will use the nominal output $Reg^m_{out\_NOM}$ as source so that register Reg^{m+1} will not receive new data and still acknowledge the last consistent data. Thus, $Done^m_{RED}$ will be unequal to $Pass^m_{RED}$.

A faulty $Pass$ signal at a redundant register would block the redundant handshake but not lead to a deadlock in the default configuration of the pipeline, i.e. remain as a latent fault. This issue is described in section 4.4.6.

4.4.5 Observable Symptoms

From the three different deadlock conditions unique symptoms for each type of fault can be derived. The common symptom for all faults in the nominal path is that $Done \neq Pass$ at a redundant and/or a nominal register. Note that all subsequently defined symptoms are described for a fault at a nominal signal, corresponding to the first fault in the default configuration. The same symptoms, however, can also be defined for faults at redundant signals. For simplicity these rules are not listed below. Latent faults at redundant signals are treated in section 4.4.6.

Symptom 1 $\left.\begin{array}{c} \overline{SHC^{m+1}_{in\_NOM}} \\ \overline{Reg^m_{in\_NOM}} \\ \overline{Ack^{mk}_{NOM}}, k = m+1 \end{array}\right\} \to Done^m_{RED} \neq Pass^m_{RED}$

Note: For a faulty nominal SHC input also $Done^m_{NOM} \neq Pass^m_{NOM}$ can be observed, but this is not relevant for identifying the fault location.

4.4. Fault Diagnosis in a Pipeline

Depending on the fault either the preceding register (SHC input faulty) or the affected register (input of register or Pass signal) will show this condition, which obviously can be used to determine the fault location. We designate this register Reg^{FL}.

It can further be seen that for SHC and acknowledge faults the input phases of Reg^{FL}_{NOM} and Reg^{FL}_{RED} are equal, while for faults at register inputs the phases are unequal:

Symptom 2 $\left. \begin{array}{c} \overline{SHC^{m+1}_{in\_NOM}} \\ \overline{Ack^{mk}_{NOM}}, k = m+1 \end{array} \right\} \rightarrow \varphi^m_{in\_NOM} = \varphi^m_{in\_RED}$

Symptom 3 $\overline{Reg^m_{in}} \rightarrow \varphi^m_{in\_NOM} \neq \varphi^m_{in\_RED}$

As the source buffer can be seen similar to an FSL register, the same observations can be made at the input to the pipeline (equivalent to the output of the source buffer): For faults at SHC inputs the phases of the nominal and redundant tokens of the source buffer will be equal, while for acknowledge and register faults it will be different. These symptoms are a special case of the latter three ones and are thus not listed explicitly here. They are used for the reconfiguration based on the "global view" of the pipeline, which means that only the symptoms at the first register of the pipeline are used to determine the appropriate reconfiguration pattern (see RU_C in section 5.4.2).

For a fault at the SHC input the input phase of Reg^{FL} is equal to the input phase of Reg^{FL+1}. In fact, Reg^{FL} already has the next value at its input. However, due to the alternating phases every second phase value is equal ($\varphi_n = \varphi_{n+2}$).

Symptom 4 $\overline{SHC^{m+1}_{in\_NOM}} \rightarrow \varphi^m_{in\_NOM} = \varphi^{m+1}_{in\_NOM}$

For faulty Pass signals the input phase of Reg^{FL} is unequal to the input phase of Reg^{FL+1}.

Symptom 5 $\overline{Ack^{mk}_{NOM}} \rightarrow \varphi^m_{in\_NOM} \neq \varphi^k_{in\_NOM}$, with $k = m+1$

Apart from these basic symptoms, some cases require exceptional handling.

Since there is no register preceding the first SHC in the pipeline, symptom 1 cannot be observed if the input data of the first SHC is faulty. In this case we can assume that if the pipeline is in a deadlock condition and there is no register where $Done^m \neq Pass^m$, the input of the first SHC in the pipeline must be faulty.

Symptom 6 $Deadlock \land (Done^m = Pass^m) \mid m = 1..m_{max} \rightarrow \overline{SHC^1_{in\_NOM}}$

If the *Pass* signal of the last register in the pipeline is faulty, symptom 5 cannot be observed because the sink does not provide the status of the input phase detector and a *Done* signal like a register. So, if a deadlock has been detected and $Done^{m_{max}} \neq Pass^{m_{max}}$, the acknowledge between the last register in the pipeline and the sink is faulty.

Symptom 7 $Done^{m_{max}} \neq Pass^{m_{max}} \rightarrow \overline{Ack^{m_{max}sink}}$

4. Self-Healing Approach

	$Reg^{m_{max}-1}$				$Reg^{m_{max}}$				Sink	
	φ_{in}	φ_{out}	$Done$	$Pass$	φ_{in}	φ_{out}	$Done$	$Pass$	φ_{in}	$Done$
NOM	φ_{n+2}	φ_{n+1}	φ_{n+1}	φ_n	X (φ_n)	φ_n	φ_n	φ_n	φ_n	φ_n
RED	φ_{n+2}	φ_{n+2}	φ_{n+2}	φ_{n+1}	φ_{n+1}	φ_{n+1}	φ_{n+1}	φ_{n+1}	φ_{n+1}	φ_{n+1}

Table 4.5: Pipeline Deadlock State for Fault at Last Register Input

Assuming two independent sink memories for the nominal and redundant pipeline path which store the data each time the phase changes, a fault at the input of the last register in the pipeline looks like symptom 4. Thus, additional symptoms have to be considered.

Due to a fault at the input of the last nominal register in the pipeline, this register will not update the output. The redundant register will receive the correct value and so the redundant sink can store the data. The resulting pipeline status is shown in Table 4.5.

This condition looks similar to a fault at a SHC input or an acknowledge fault (symptom 2). The difference is that for a fault at the input of the last register, the nominal and redundant input phases of the last register are different, whereas for the other mentioned fault conditions the input phases are equal. Therefore the following additional rules were defined.

Symptom 8 $\left. \begin{array}{c} \overline{SHC_{in\_NOM}^{m+1}} \\ \overline{Ack_{NOM}^{mk}}, k = m+1 \end{array} \right\} \rightarrow \varphi_{in\_NOM}^{m+1} = \varphi_{in\_RED}^{m+1}$

Symptom 9 $\overline{Reg_{in\_NOM}^{m_{max}}} \rightarrow \varphi_{in\_NOM}^{m_{max}} \neq \varphi_{in\_RED}^{m_{max}}$

A fault within a SHC will prevent the SHC from updating the output due to inconsistent data at the input of the FSL combinational logic inside the SHC. Depending on the fault location and the SHC configuration, either the nominal or the redundant output or both will not change value. If both outputs stop, the behavior follows the rules described above for SHC faults. If only one output is affected, the subsequent register using the respective nominal or redundant value will not get new data, i.e. the input phase detector will not change state, while the other register receives and processes a new value. This behavior is equal to a fault at a register input. Thus, by observing the symptoms defined above, a fault within a SHC might not be distinguishable from a fault at the input of the subsequent register. The handling of this particular condition must be considered in the reconfiguration unit, e.g. by always first performing a SHC reconfiguration and moving on to the corresponding reconfiguration for register faults if the SHC reconfiguration was not sufficient to solve the problem. This approach is also applicable for fine-granular SHC implementations, see section 4.5.2.

Table 4.6 summarizes the symptoms, and shows which combination of symptoms identifies the different faults. All symptoms marked with "X" must occur in order to clearly identify the corresponding fault.

Figure 4.12 shows an exemplary algorithm using the above symptoms in a graphical representation. The reference to the symptoms is given in brackets.

4.4. Fault Diagnosis in a Pipeline

Symptoms	General			Exceptions		
	SHC_{in}	Reg_{in}	Ack	SHC_{in}^1	$Reg_{in}^{m_{max}}$	$Ack^{m_{max}sink}$
1	X	X	X		X	X
2	X		X		X	
3		X				
4	X					
5			X			
6				X		
7						X
8	X		X			
9					X	

Table 4.6: Summary of Symptoms due to Single-Faults in the Pipeline

4. Self-Healing Approach

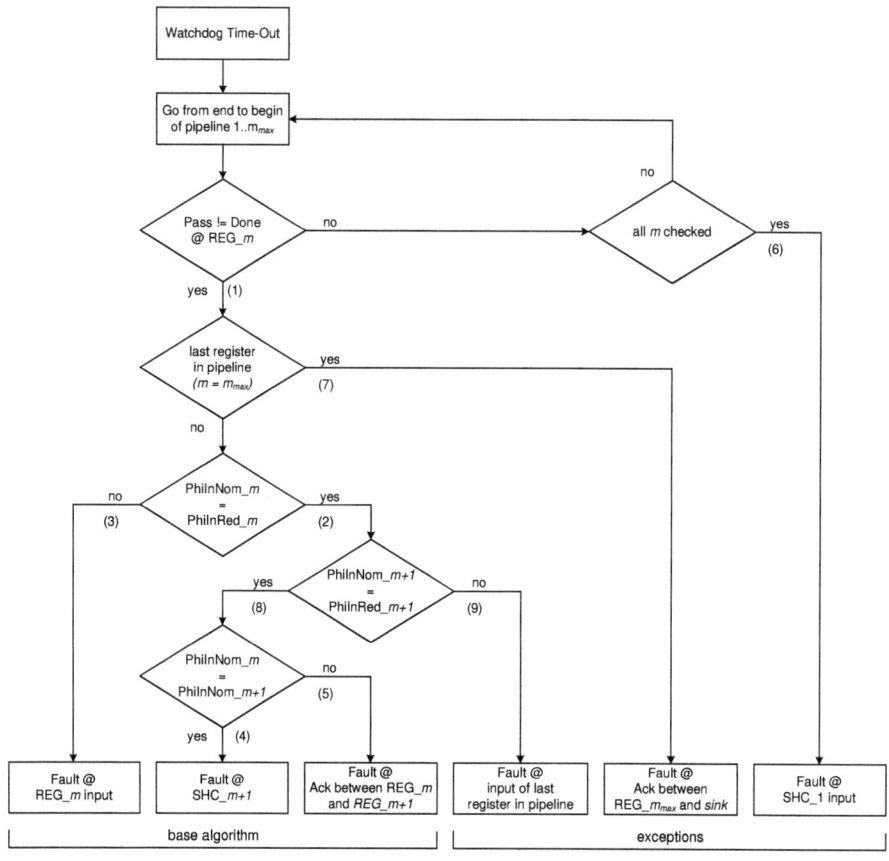

Figure 4.12: Reconfiguration Rules and Algorithm for Local Observability

4.4.6 Effects due to Multiple Faults

In this section the conditions for multiple faults are evaluated. As we expect the probability of multiple permanent faults occurring at the same time (or in close temporal proximity) to be negligible, we rather consider here the case of a single permanent fault emerging after a finished reconfiguration due to a previous fault.

Multiple Faults at SHC Inputs

For the first fault at a SHC input a reconfiguration of only the affected SHC is sufficient. The fault can be bypassed locally inside the SHC, and - if a configuration is applied, so that both outputs are correct - fault-free conditions prior and after the defective SHC are established. Any additional fault in the pipeline can thus be identified following the initial diagnosis and rules.

Multiple Faults at Register Inputs

Without acknowledge switches a fault at a register input can only be repaired by separating the two pipeline paths at least up to the defective register (see section 4.3.3). With the simple reconfiguration any additional fault in the same pipeline path will therefore not affect the functionality, because this pipeline path is deactivated anyway. An additional fault in the other path cannot be repaired. For the optimized reconfiguration these statements are true for faults prior the defective register. After the SHC located subsequently to the defective register, nominal conditions are established again. Thus, the initial rules are valid for faults occurring at least one SHC after the defective register (see Figure 4.19c).

With acknowledge switches the fault can be bypassed locally by routing both the acknowledge and data signals via the redundant pipeline path. In appropriate distance from this faulty register (at least one SHC in between, see Figure 4.19d) the pipeline will then look fault-free, and all the symptoms described previously are valid. Within a distance of one register and one acknowledge switch the symptoms depend on the previously performed reconfigurations and can differ from the initial ones. In this case the resulting symptoms could lead to a wrong reconfiguration, i.e. the fault situation cannot be repaired.

Multiple Faults at Acknowledge Signals

Faults at acknowledge signals cause the same behavior as faults at register inputs, thus the same effects as described above can be observed.

Fault Combinations on SHC and Register Inputs

For the fault combination of SHC input and register input the sequence of the fault occurrence will influence the behavior of the pipeline and the success of the reconfiguration.

Without acknowledge switches:

- If the fault at the register input occurs first, the pipeline paths need to be separated. Any further fault at the same pipeline path will thus not affect the functionality.

- If the fault at the SHC input happens first, the reconfiguration will re-establish the fault-free conditions due to the local bypass. The subsequent fault at the register input will cause the described symptoms for register faults, and thus the pipeline paths will be separated.

- If the defective register is located later in the pipeline than the defective SHC only the redundant path will provide valid results (Figure 4.13a). Otherwise, both the nominal and redundant outputs of the pipeline will be correct (Figure 4.13b).

With acknowledge switches basically the same statement is valid, as well as the statement for multiple faults at register inputs. The only exception is that the pipeline paths will not be separated.

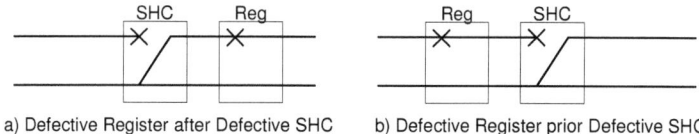

Figure 4.13: Multiple Faults at SHC and Register Inputs

Fault Combinations on SHC Inputs and Acknowledge Signals

Since faults at acknowledge signals have the same effect as faults at register inputs, the same statements as above are valid.

Fault Combinations on Register Inputs and Acknowledge Signals

This combination causes the same behavior as multiple faults at register inputs.

4.4.7 Summary

Each of the three single-fault conditions "faulty SHC input", "faulty register input" and "faulty acknowledge signal" can be clearly identified with the symptoms presented above. Depending on the location of the fault within the pipeline, different symptoms will appear. Three basic locations have to be distinguished, which are "begin of the pipeline", "end of the pipeline" and "between begin and end of the pipeline" and need to be treated differently.

In general, after a successful repair, the pipeline with acknowledge switches behaves like in the fault-free configuration and causes the same symptoms in a new fault situation. However, there exist some particular cases, where the application of the "single fault diagnosis" to multiple faults can lead to a wrong identification and thus to an incorrect reconfiguration (see section 5.4). This is an issue for the design of the reconfiguration unit, which needs to be intelligent enough to apply different rules, depending on the history of the circuit.

4.5 Reconfiguration Rules and Algorithm Efficiency

4.5.1 Reconfiguration on Pipeline-Level

In order to evaluate the effectiveness of different reconfiguration algorithms applied on a circuit, first the number of possible circuit configurations leading to a successful repair must be determined. As already explained in section 4.3, all single faults can be tolerated by the architecture. This is not surprising considering the duplication of the pipeline. The strength of the fine-grained self-repair becomes visible, however, in the presence of multiple faults.

Table 4.7 presents formulas to calculate the number of non-repairable double faults as a function of the number n of pipeline stages. The correctness of these formulas was verified by simulations. The criterion for a successful repair was, that the configuration creates a circuit, where a data and acknowledge path exists, so that at least one of the two outputs delivers correct results, and the pipeline does not end up in a deadlock.

Note that the formulas for the case "without AS" assume the "simple" reconfiguration by separating the pipeline paths as described in section 4.3.3. This algorithm has been used for the hardware injection experiments with RU_C (see section 5.4.2).

faulty signals	number
without AS	
$SHC^j_{In\_Nom} \wedge SHC^j_{In\_Red}, j = \{1,...,n\}$	n
$Reg^j_{In\_Nom} \wedge Reg^k_{In\_Red}, j,k = \{1,...,n\}$	n^2
$Pass^j_{Nom} \wedge Pass^k_{Red}, j,k = \{1,...,n\}$	n^2
$Reg^j_{In\_Nom|Red} \wedge Pass^k_{Red|Nom}, j,k = \{1,...,n\}$	$2*n^2$
$Reg^j_{In\_Nom|Red} \wedge SHC^k_{In\_Red|Nom}, j,k = \{1,...,n\}$ exception: $\mid j-k \mid \geq 1$	$2*(n^2 - \sum_{i=2}^{n-1}(n-i))$
$Ack^j_{Nom|Red} \wedge SHC^k_{In\_Red|Nom}, j,k = \{1,...,n\}$ exception: $\mid j-k \mid \geq 1$	$2*(n^2 - \sum_{i=2}^{n-1}(n-i))$
total	$8*n^2 + n - 4*\sum_{i=2}^{n-1}(n-i)$
with AS	
$Pass^n_{Nom} \wedge Pass^n_{Red}$	n
$SHC^n_{In\_Nom} \wedge SHC^n_{In\_Red}$	n
$Reg^n_{In\_Nom} \wedge Reg^n_{In\_Red}$	n
$Pass^n_{Nom|Red} \wedge Reg^n_{In\_Red|Nom}$	$2*n$
$SHC^n_{In\_Nom|Red} \wedge Pass^{n-1}_{Red|Nom}$	$2*(n-1)$
$SHC^n_{In\_Nom|Red} \wedge Reg^{n-1}_{In\_Red|Nom}$	$2*(n-1)$
total	$9*n - 4$

Table 4.7: Number of Non-Repairable 2-Fault Configurations in an n-Stage Pipeline

Basically, in both variants (with and without acknowledge switches) all fault combinations, where the nominal *and* redundant element of the same pipeline stage are affected, cannot be repaired (e.g. nominal and redundant register input). Without using acknowledge switches also all fault combinations affecting any nominal and any additional redundant register input or acknowledge signal cannot be repaired. Finally, there are some specific combinations of faults that cannot be repaired:

- Without using acknowledge switches all combinations of a faulty SHC input and a faulty register input or acknowledge signal in the other path cannot be repaired $(2*n^2)$, except those combinations where the faulty register or acknowledge signal is located earlier in the pipeline than the SHC fault, and there is at least one SHC in between $(2*\sum_{i=2}^{n-1}(n-i))$. The reason for this exception is obvious: Let's assume a faulty nominal SHC input, which requires to configure the SHC to use the redundant input. If there occurs another fault at a register input or acknowledge signal later in the redundant pipeline path (Figure 4.14a), the redundant handshake is blocked and thus the register providing the data for the redundant SHC input will not be updated any more.

- If the additional fault occurs earlier in the pipeline, the redundant handshake is only blocked up to the faulty register, and the redundant SHC input can receive new data (Figure 4.14b) - provided, that there is at least one SHC in between. Otherwise the register providing data to the redundant SHC input would also be defective (Figure 4.14c).

- With acknowledge switches these cases can be repaired, except if there is a fault at a SHC input and at a register input or acknowledge signal in the previous stage in the other path (e.g. Figure 4.14c).

The difference of the two pipeline architectures with respect to fault tolerance is clearly visible in Figure 4.15. The relative improvement of the success rate with rising n can be explained by the fact, that with longer pipelines the number of fault combinations where no neighbor stages are affected is increasing faster than the number of troublesome situations described above.

For the evaluation whether a valid path exists in the pipeline for a particular fault situation, all faults can be injected simultaneously, because all possible paths (pipeline configurations) are checked until a valid data *and* acknowledge path is established, without any dependency on previous reconfigurations, i.e. assuming a "perfect" algorithm.

The number of fault combinations equals $\binom{FPos}{NofFaults}$, where $FPos = 6*n$ (three fault locations per pipeline path and stage), and n =number of pipeline stages. Table 4.8 presents the results for a 5-stage pipeline ($FPos = 30$) and 1 to 3 faults. The same circuit was used for the hardware experiments described in section 5.5. These numbers show the theoretical best case and can be used as benchmark to evaluate the efficiency of a reconfiguration algorithm. Without any fault all possible configurations are valid. For more than two faults no formulas were developed, instead simulations covering all possible fault combinations and pipeline configurations have been performed to determine the maximum number of successful reconfigurations.

4.5.2 Embedded Fine-Granular SHC Reconfiguration

Faults inside a SHC will prevent the SHC from updating either one or both outputs due to the inconsistent data on any internal FSL logic input. Thus, the symptoms of either a fault at a SHC input (both outputs blocked), or of a fault at the input of the subsequent register (one SHC output blocked), can be observed (see section 4.4.5).

So far, the reconfigurations considered coarse granular SHCs only, i.e. the implemented logic can either use the nominal or the redundant input and the SHC will produce two respective outputs.

4.5. Reconfiguration Rules and Algorithm Efficiency

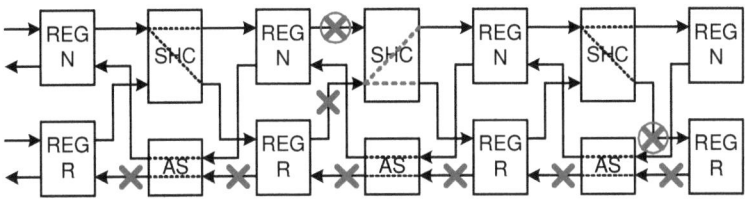

a) Un-Repairable Double Fault Combination SHC-Reg_In w/o AS Reconfiguration

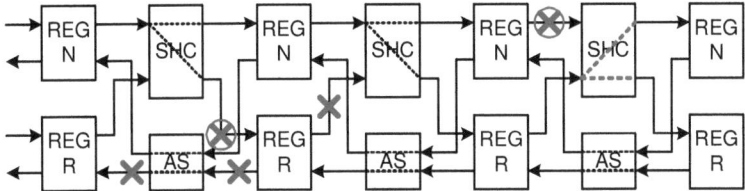

b) Successful Repair of Double Fault Combination SHC-Reg_In

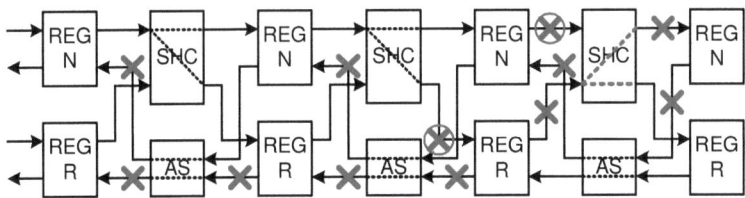

c) Un-Repairable Double Fault Combination SHC-Reg_In

Figure 4.14: Fault Combinations of SHC and Reg Faults

No. of faults	fault combinations	max	success
without AS			
0	0	-	100.0%
1	30	30	100.0%
2	435	254	58.4%
3	4060	1126	27.7%
with AS			
0	0	-	100.0%
1	30	30	100.0%
2	435	394	90.6%
3	4060	2988	73.6%

Table 4.8: Theoretical Maximum of Successful Reconfiguration Paths in 5-Stage Pipeline

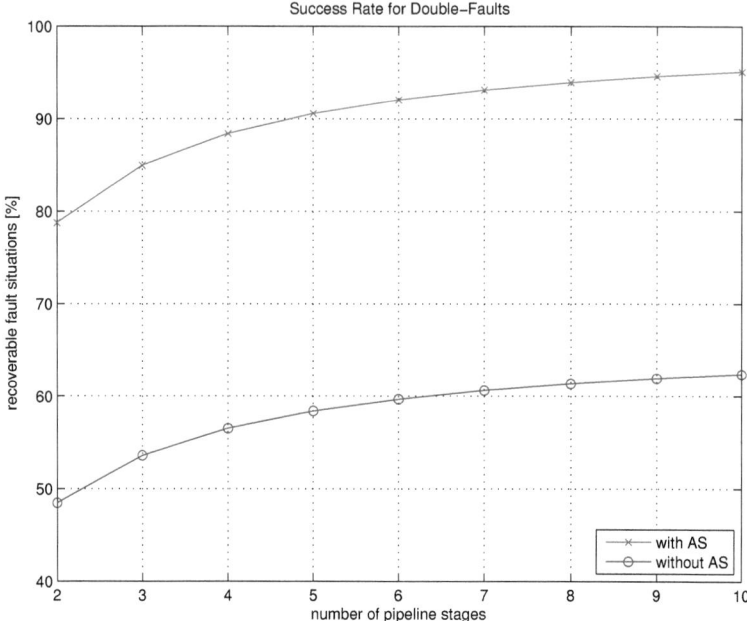

Figure 4.15: Comparison of Double-Fault Handling

Following the described symptoms, an internal fault affecting only one output would be handled as a register fault, which is indeed the only possibility for coarse granular SHCs.

With fine granular SHCs some new topics come up:

1. Fine granular self-healing logic is built from several SHCs, i.e. several reconfiguration inputs have to be handled.

2. For a configuration to "completely redundant" like for a coarse granular logic, several reconfiguration inputs have to be set correctly.

3. Although an internal fault can cause symptoms like a fault at a register input, it can be repaired within the SHC. A complete switch to the redundant part would be possible, but is a waste of resources.

This means that faults being identified as faults at a register input should be handled as follows. First, *only the SHC* preceding the suspicious register should be reconfigured. This can be done with random configuration patters, as the handshake is not affected and no unintended data capturing will occur. If the deadlock cannot be removed this way, it can be concluded that the fault is located either late in the SHC so that it cannot be removed by SHC reconfiguration, or it is located at the input of the subsequent register. In both cases it has to be treated as register fault.

This 2-level reconfiguration has not been considered in the simulations and experiments within this thesis. However, SHC and pipeline reconfigurations have been investigated in detail as separate topics (see sections 5.3 and 5.5).

4.6 Overhead of the Self-Healing Approach

This section presents an analysis of the overhead the self-healing approach introduced in terms of logic resources compared to a standard FSL circuit. Only the overhead caused by the instrumentation of the application circuit is analyzed. The overhead of the reconfiguration unit depends significantly on the algorithm and thus cannot be calculated in a general way.

The overhead in terms of logic resources, which the self-healing approach adds, accumulates from (i) the redundant logic and (ii) the control logic needed to operate the redundant logic (routing elements, i.e. switches).

The overhead was analyzed at the example of an adder. In the fine granular (FG) implementation each gate of the adder is a basic SHC (AND, OR, etc.) with two inputs, which provides much more routing possibilities. In contrast, the coarse granular (CG) adder is built as one single SHC which contains two equal adder circuits. The principle of fine and coarse granular SHC is depicted in Figure 4.4.

Independent from the SHC granularity, the logic resources in a SHC will be doubled compared to a standard FSL implementation. The number of switches (one for each input) and the number of reconfiguration inputs (two for each SHC), however, makes a significant difference. The resource occupation for a two input logic function (34 equivalent logic gates) and a switch (12 equivalent logic gates) was determined from the Xilinx synthesis report.

Fig. 4.16 shows the resource overhead (OH) compared to a standard FSL implementation versus the width of the adder. The relative resource overhead for the FG SHC is constant (factor of 3.4), because each bit of the adder requires the same amount of resources (1-bit adder). For the CG SHC adder the overhead decreases from approximately 2.35 for a 2-bit adder to 2.15 for a 10-bit adder because the constant overhead of the switches decreases relative to the total amount of resources.

For complex circuits a completely fine granular implementation is not meaningful due to the high number of reconfiguration inputs that have to be controlled by the reconfiguration unit. As shown in the figure, for a 10-bit adder more than 10000 reconfiguration inputs would be needed. In this case, the resource overhead of the reconfiguration unit would very likely dominate the total amount of resources and keep the overhead of the SHC implementation insignificant. A fine granular implementation is thus only reasonable for small applications that are particularly critical or where an extremely high failure rate is expected. Of course, fine and coarse granular circuits can be combined as needed in the application to achieve the reliability goal. A comparison of the two implementations with respect to fault tolerance is presented in section 5.3.1.

In addition to the redundant logic in the SHCs the self-healing concept doubles the resources of the pipeline registers due to the redundant pipeline path. However, it is expected that in a typical application the amount of resources used for the logic (implemented in SHCs) will be

4. Self-Healing Approach

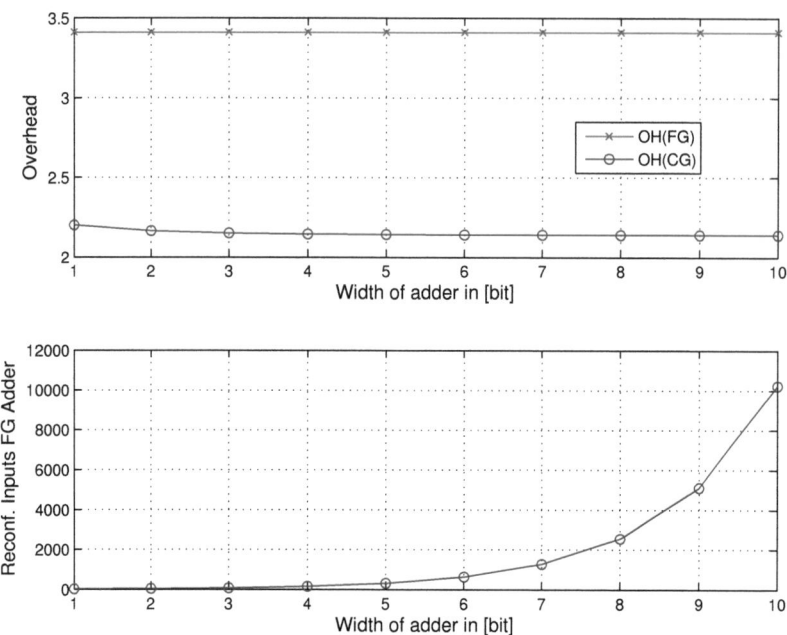

Figure 4.16: Overhead and Number of Reconfiguration Inputs vs. Width of a Full-Adder

much higher than for the "administrative logic" in the pipeline, so that the overhead of the SHCs dominates. Figure 4.17 shows the total overhead of the self-healing approach (without reconfiguration unit) versus the ratio of resources related to the registers (CG SHC is considered with factor 2.15). For a circuit consisting only of registers (100%) the overhead settles at a factor of 2 which is caused by the redundant pipeline.

4.6. Overhead of the Self-Healing Approach

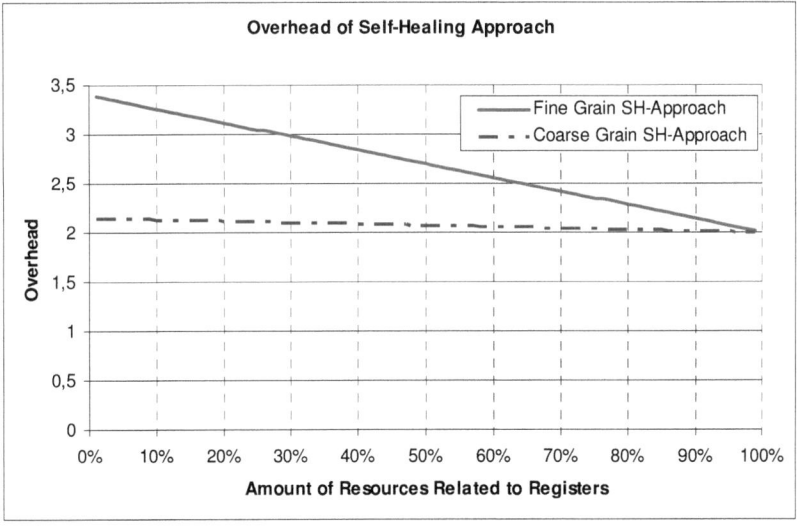

Figure 4.17: Overhead of Self-Healing Approach

4.7 Annex

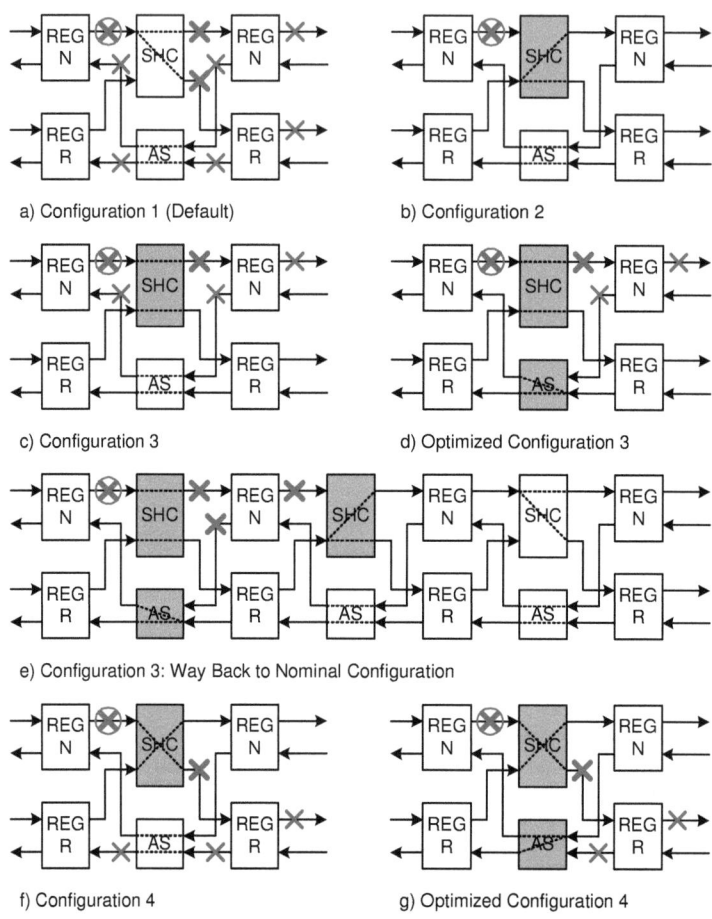

Figure 4.18: Fault at SHC Input

4.7. Annex

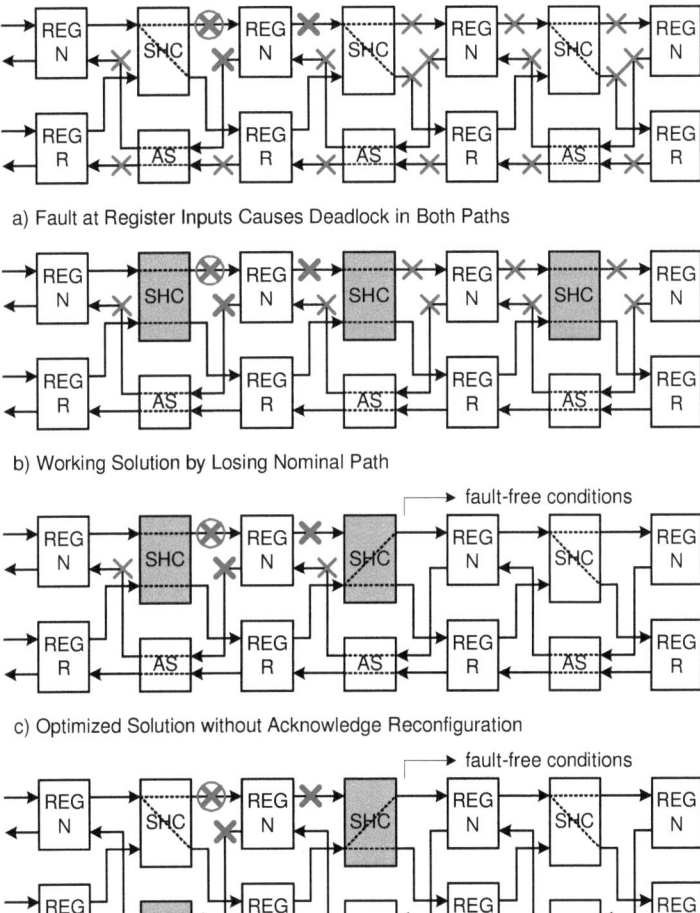

Figure 4.19: Fault at Register Input

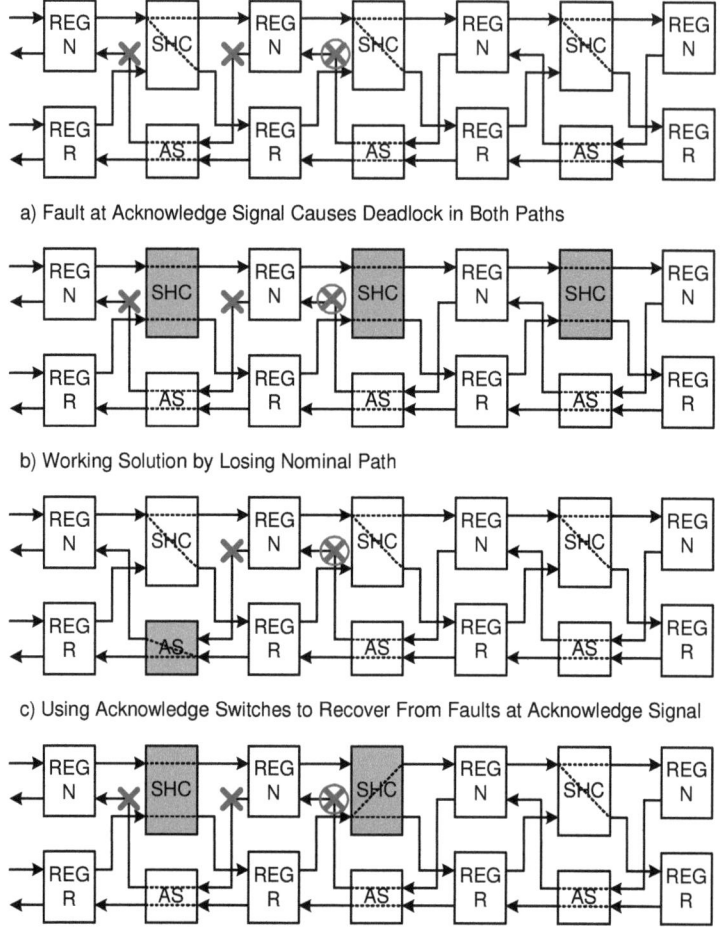

Figure 4.20: Fault at Acknowledge Signal

Chapter 5

Analysis, Simulations and Experimental Results

5.1 Introduction

This chapter summarizes all results of the steps performed in scope of the thesis. Table 5.1 lists the activities and provides a reference to the relevant section where they are described in detail.

Description	Method	Reference
SHC Related Activities		
SHC Fault Tolerance, Fine vs. Coarse Granular	Simulation	§5.3.1
SHC Optimization	Analysis	§5.3.2
Pipeline Related Activities		
Elaboration of Pipeline Reconfiguration Algorithm	Simulation	§5.4
Verification of Pipeline Reconfiguration with H/W Fault Injection	Experiment	§5.5
Architecture Related Activities		
Show Case: Complex Self-Healing Circuit	Experiment	§5.6
Reliability Analysis of Self-Healing Pipeline	Analysis	§5.7

Table 5.1: Summary of Simulations, Analysis and Experiments

The explanations in the previous chapter already showed that the complexity of the whole self-healing approach is rather high. It was quite clear that it is not possible to use one particular model or environment for all investigations. In particular the focus was completely different between e.g. the determination of fault tolerance in a SHC and the elaboration of reconfiguration strategies for the pipeline. Therefore a combination of analysis, simulations (on different abstraction levels) and hardware experiments has been chosen to validate the self-healing approach. Each model was optimized for a particular goal and, as far as feasible, was verified by supplementary experiments.

For example, exhaustive fault injection simulations with Modelsim would not have been feasible due to the large simulation time, so an abstract Matlab model neglecting the timing was developed. The same fault injections have been performed in hardware, which on the one hand verified the correctness of the abstract simulation model, and on the other hand proved the concept under real conditions.

5.2 Environment

5.2.1 General

This section describes the major simulation and hardware environments developed and used in scope of this thesis.

The environments used to investigate the pipeline configurations, both for simulations and hardware experiments, are described in detail below. One important aim was the ability to emulate the same situations in both the simulation model and the hardware environment on different abstraction levels and with different observability (level of insight and access to internal circuit parameters). This was supported by having a compatible interface for the stimulus definition and fault injection. With the simulation and hardware environment described herein it is possible to perform both simulation-based and hardware fault injection.

For the exploration of SHCs no such sophisticated environment was necessary, so no general description is given here. The details about the chosen exemplary circuits that were implemented and evaluated can be found in the respective section.

5.2.2 Simulation Environment for Pipeline Reconfiguration

The model used to evaluate pipeline reconfigurations and different algorithms was established in Matlab. It considers basic behavioral rules of the involved elements (e.g. the conditions when a FSL register is transparent) and then calculates the new values and phases. Signal vectors are modeled in an abstract way as a value that can be either *correct* or *incorrect*. Fault injections on rail-level are not possible, neither is the simulation of circuit timings. This considerably improves the simulation performance and makes the model insensitive to hazard-based effects as described in section 2.4.5. Timing effects due to crossing of nominal and redundant data and acknowledge paths (see section 4.3.5) are neglected as well here for a first estimate.

The model is composed from two main blocks:

- Target circuit: is designed from several *elements* such as SHC, register, acknowledge switch, and builds the *pipeline*, which is investigated in various fault conditions and with different reconfiguration algorithms

- Reconfiguration Unit: simulates the functionality of a reconfiguration unit, including the watchdog circuit and the reconfiguration controller. Different reconfiguration algorithms can be configured for the reconfiguration controller.

Target Circuit

The target circuit for the investigation of pipeline reconfigurations shall on the one hand be as small as possible, so that simulations and experiments can be performed in a reasonable time, but it shall also be complex and long enough to get rid of "border effects". Thus, a 5-stage pipeline is used as target circuit with combinational logic implemented as SHCs in between. Two pipelines were investigated, one only with SHCs as reconfigurable elements (Figure 5.1)

5.2. Environment

and a second one using also acknowledge switches (Figure 5.2). The default configuration for the SHCs is that the nominal input is used as source for both the nominal and redundant logic. For the acknowledge switches two variants have been investigated: (i) in the default configuration the nominal *Done* is routed to the nominal *Pass* and the redundant *Done* to the redundant *Pass* of the preceding registers (as it is the case without acknowledge switches) and in (ii) both the nominal and redundant registers receive the nominal *Pass* signal by default, i.e. the handshake path follows the data path.

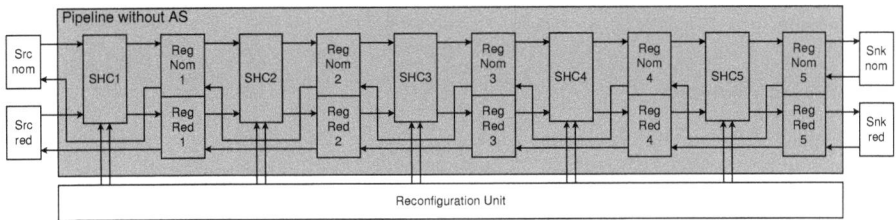

Figure 5.1: Simulated Pipeline without Acknowledge Switches

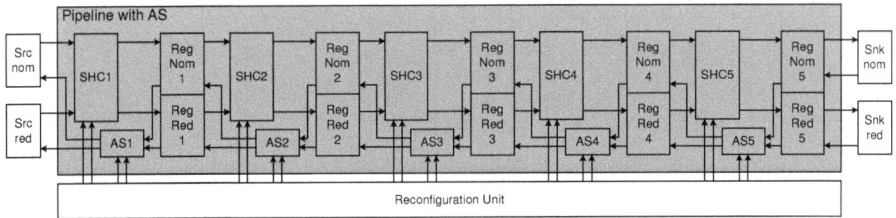

Figure 5.2: Simulated Pipeline with Acknowledge Switches

Two independent sources and sinks are used. Each calculation cycle defines the new value and phase independently for each element and so the new conditions/tokens move through the pipeline. The sources deliver values from 0 to 9 in alternating phases each time the acknowledge input changes state and wrap around after the value 9 has been reached.

Fault Injection

Faults can be injected at any time on register inputs, *Pass* signals and SHC inputs by simply replacing the value property of a signal with an 'X', indicating an inconsistent or incorrect value, respectively. These signals are then not evaluated any more for the determination of a token transfer or capturing (the rules can be seen in listing 5.7, 5.8 and 5.9). It is possible to inject any number of faults at a time.

5. Analysis, Simulations and Experimental Results

Reconfiguration Unit

The model of the reconfiguration unit defines the rules for the applied reconfiguration. If there is no change in data and phases in the pipeline after a calculation cycle, a deadlock is assumed. Depending on the pipeline length, several calculation cycles need to be performed in order to allow the fault to settle and cause a deadlock.

If a deadlock is detected, the reconfiguration unit checks the internal pipeline states (phases and values) according to the symptoms elaborated in section 4.4.5, and performs defined actions accordingly. Different reconfiguration algorithms can be easily tried the same target circuit for comparison.

Simulation Elements

2-Input SHC The SHC is modeled as a simple feed-through (i.e. no combinational function implemented) with two reconfiguration inputs. Each input/output has a value and a phase property (Figure 5.3). The model assumes a correctly working combinational logic, i.e. a defective input (causing inconsistent data) does not propagate to the SHC output, as it will be blocked by the FSL logic (see chapter 2.4.2). Internal faults cannot be injected, but as outlined in section 4.5.2 they will cause the same behavior as faults at SHC or at register inputs.

The rules of the SHC model are presented in listing 5.7.

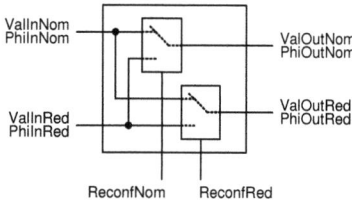

Figure 5.3: Properties of SHC Simulation Model

Register The register is modeled as a redundant component, i.e. two independent registers are considered. Apart from the data inputs/outputs acknowledge signals are needed as well (Figure 5.4). The model assumes a correctly working register control logic, i.e. a defective input (meaning inconsistent) does not propagate to the register output (see chapter 2.4.3). A fault in the control logic would cause an incorrect output or the *Done* signal to be not updated, which can be modeled as a fault on the subsequent SHC or sink input, or as a fault on the *Pass* input of the preceding register.

The rules of the register model are presented in listing 5.8.

Acknowledge Switch The acknowledge switch is basically a switch matrix which routes the acknowledge signals according to the configuration inputs (Figure 5.5).

The rules of the acknowledge switch model are presented in listing 5.9.

5.2. Environment

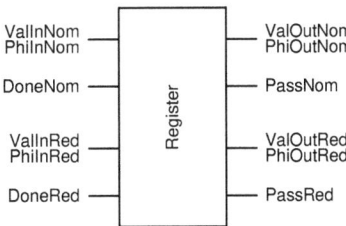

Figure 5.4: Properties of Register Simulation Model

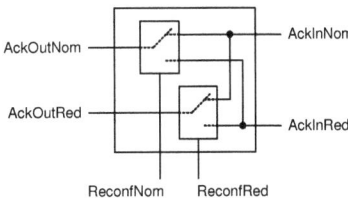

Figure 5.5: Properties of Acknowledge Switch Simulation Model

5.2.3 Environment for Hardware Experiments

The hardware fault injection experiments were performed using the Memec Virtex-4 FX12 LC evaluation board. All activities, like defining the source data, applying faults, reading data from the sink buffer, etc., can be controlled via an USB interface. Basically a simple terminal program would be sufficient, but for convenience a dedicated application with a graphical user interface was designed. A screen snapshot can be found in Figure 5.29 at the end of this chapter. This application includes a simple script interpreter which eases the definition of test sequences. Listing 5.6 presents an example of such a scenario.

For the fault injection *saboteur* components were designed, which allow to apply stuck-at-0 and stuck-at-1 faults to any signal in the circuit. Figure 5.6 shows the component and its logic table.

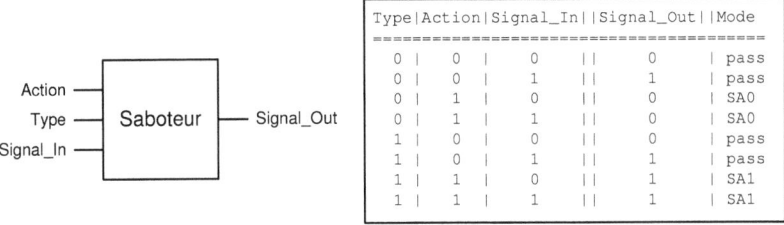

Figure 5.6: Saboteur Component

5. Analysis, Simulations and Experimental Results

For fault injection experiments the saboteur must be placed manually in the design on any signal that shall be able to be modified. The principle environment and examples for a saboteur application are given in Figure 5.7.

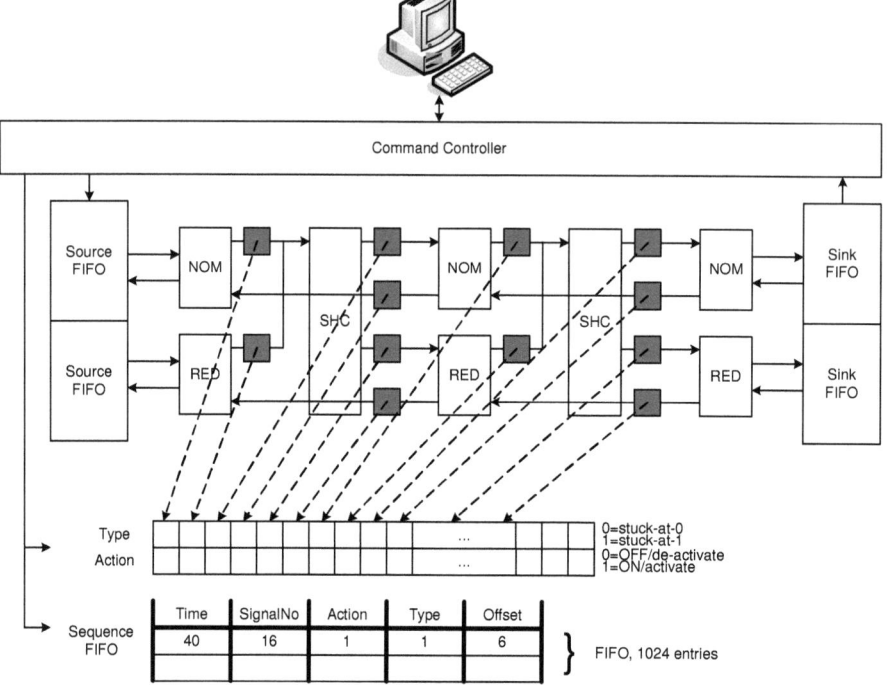

Figure 5.7: Fault Injection Environment

Each saboteur receives the target signal and two configuration signals as input. The configuration signals allow defining the type of fault (stuck-at-0, stuck-at-1) and control the activation of this fault injection. The saboteur configuration signals are mapped to a distinct bit position in a "Type" and an "Action" register, which are controlled from the fault injector. Each bit position refers to exactly one saboteur, i.e. one signal. The fault injector processes a configuration table which must be pre-loaded into a FIFO buffer before starting the application. Each entry in the FIFO contains

- The time when the fault injection shall take place, i.e. the number of clock ticks of a dedicated fault-injector clock. In the experiments described herein a pipeline handshake signal was used as clock, so that the time of fault injection was equal to the number of handshake cycles in the pipeline (corresponding to the number of processed tokens).

- The signal number to which the fault shall be applied (corresponding to the bit offset in the "Type" and "Action" register).

5.3. Reconfiguration of Self-Healing Cells

- The action to be performed (activate, de-activate the fault).
- The fault type (stuck-at-0, stuck-at-1).
- An "Offset" field, which defines the delay in number of system clock cycles after the fault injector is triggered. This allows to shift the actual fault injection in fine time steps within the handshake cycle.

After the fault injector is started, it waits for the next trigger condition and then performs the configured action. If the time is set to zero the next injection is performed immediately without waiting for the next handshake cycle. This gives the possibility to apply multiple faults nearly simultaneously.

Furthermore, the fault injector allows to read-out the actual reconfiguration vector of the application, and also to manually define a reconfiguration vector for debugging purposes.

5.3 Reconfiguration of Self-Healing Cells

5.3.1 Fault Tolerance of Fine and Coarse Granular Self-Healing Cells

Introduction

In this section the tolerance of fine and coarse granular SHC circuits against single and multiple faults is compared, as well as the difference of the occupied resources. The goal of the simulation was to prove the anticipation that for any single stuck-at fault a working configuration can be found, and that the SHC is also highly tolerant against multiple faults.

Simulation Description

The target circuit was a full adder which was designed in two ways:

1. The first circuit (circuit A) was built from low level SHCs (Figure 4.4a; example given for a half adder in Figure 4.4b).
2. In the second circuit (circuit B, Figure 4.4c) two complete full adders were implemented in the SHC, which basically act as active-redundant circuits.

Both circuits were designed as RTL model in VHDL and Modelsim was used for the simulation. The reconfiguration inputs of the SHCs were controlled by a testbench-model of a reconfiguration unit. It was implemented as a counter, which ensures that all possible configurations are tested. The drawback is the large simulation time, so that only a 1-bit adder was simulated. However, since adders with higher bitwidths can be constructed by cascaded 1-bit adders, and faults on the external interfaces have been considered in the simulation, the results obtained from the 1-bit adder are also representative for larger adders.

A list of all internal signals and external interfaces, including the reconfiguration inputs, was extracted from the RTL design. Then up to 11 faults were injected on these signals. The first signal was chosen sequentially from the extracted list. To obtain a good statistical distribution the additional signals, as well as the fault type (stuck-at-0 or stuck-at-1), were selected randomly. The applied sequence guaranteed that each signal was subjected at least once to both stuck-at faults per simulation run.

After the fault injection all valid input stimuli were applied in alternating phases. The 1-bit full adder provides three inputs (a, b, c_{in}), resulting in 8 combinations for each phase or 16 combinations in total. Due to the storage elements in the FSL gates the circuit behavior in the presence of errors depends on the history. To consider this dependency, five independent simulation runs (with random signal selection as described above) were performed, and the mean values of the success rates were taken as the final result. In total, 22720 fault conditions with the fine granular circuit (circuit A) and 17600 for the coarse granular circuit (circuit B) were simulated[1].

In case of a deadlock the counter controlling the reconfiguration inputs started to count up. As soon as a configuration was found that resumed the circuit's operation, it was assumed that the circuit has been repaired correctly. The result of the adder (sum and carry) was compared to the expected value and the circuit was defined to be working if at least one of the two redundant outputs showed the correct result. If the applied configuration did not remove the deadlock, the reconfiguration process was continued by trying the next configuration pattern until all possible patterns were applied. If no working configuration could be found, this fault combination was declared as uncorrectable.

Results

The simulations confirmed that for both circuits it is possible to repair all single faults. For multiple faults there is a high probability for a successful repair which, however, depends on the fault location in the circuit. If e.g. two faults affect both the nominal and the redundant path within a SHC, the circuit will fail. The same applies if a signal and its associated reconfiguration input are affected by a permanent fault at the same time. The summary of the simulations is presented in Table 5.2. To obtain the resource occupation of the two circuits, both designs were synthesized into a Xilinx Virtex-4.

It can be seen, that even with 11 simultaneously injected faults, which equals about 10% of the circuit's signals being defective, still about 54% (circuit A) and 37.5% (circuit B), respectively, of the fault constellations could be repaired. The resource overhead of the fine granular circuit is approximately 40% compared to the coarse granular circuit. However, as can be seen in Figure 5.8, the gain of fault tolerance with the fine granular implementation is also significant, in particular for a small number of faults where the probability for multiple faults within the same reconfigurable element is low.

5.3.2 Optimization of Self-Healing Cells

In the following subsections three different SHC architectures, as presented in Figure 5.9, are analyzed with respect to their resource occupation, fault tolerance and the influence on the reconfiguration process.

In architecture A both redundant logic functions can use either the nominal or the redundant SHC inputs. Architecture B allows to individually select either the nominal or redundant SHC inputs for each logic function. In architecture C each operand of both redundant logic functions can be selected from either the nominal or redundant SHC inputs.

---

[1] The different number of fault conditions results from the different number of internal signals, e.g. 142 signals * 16 input combinations * 2 fault types * 5 simulation runs = 22720 fault combinations for circuit A.

5.3. Reconfiguration of Self-Healing Cells

Table 5.2: Comparison of SHCs with Different Complexity

Circuit A: 22720 fault conditions Circuit B: 17600 fault conditions	circuit A fine gran.	circuit B coarse gran.	A/B comparison
number of signals	142	110	+29.1%
number of reconfig. inputs	10	2	+400.0%
equivalent gate count	580	412	+40.8%
failed with 1 fault	0.0%	0.0%	0.0%
failed with 2 faults	2.1%	5.0%	-57.2%
failed with 3 faults	5.5%	14.3%	-61.9%
failed with 4 faults	11.0%	22.8%	-51.8%
failed with 5 faults	16.0%	30.5%	-47.4%
failed with 6 faults	22.0%	39.0%	-43.7%
failed with 7 faults	26.8%	45.2%	-40.6%
failed with 8 faults	32.5%	49.9%	-34.9%
failed with 9 faults	38.1%	54.7%	-30.4%
failed with 10 faults	42.3%	59.2%	-28.5%
failed with 11 faults	46.0%	62.5%	-26.5%

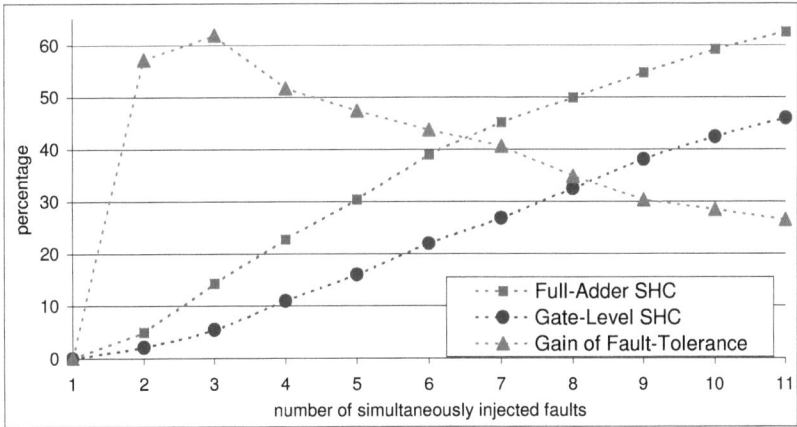

Figure 5.8: Fail-Cases and Gain of Fault-Tolerance vs. Number of Faults

The number of reconfiguration switches of the three architectures is the same, however they are controlled by a different number of reconfiguration inputs. In architecture A there will always be one, in architecture B always two reconfiguration inputs, while for architecture C the number of reconfiguration inputs depends on the number of operands fed into the SHC.

5. Analysis, Simulations and Experimental Results

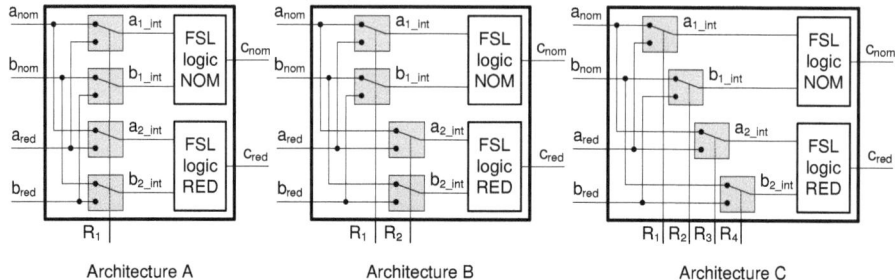

Figure 5.9: Different Self-Healing Cell Architectures

Resource Comparison

In all three architectures the number of SHC inputs and outputs is the same, and each input signal (more precisely: each rail) requires one switch. For comparison the three different SHC architectures from Figure 5.9 were synthesized into a Xilinx Virtex-4 FPGA. As the logic circuit in the SHC has no relevance for this analysis, the two internal logic inputs were simply combined to one output vector. This synthesis was performed with 2-bit, 4-bit and 8-bit wide SHC input signals. The results are summarized in Table 5.3 as absolute numbers and related to one (signal-)bit.

Table 5.3: Resource Occupation for the Different SHC Architectures

		absolute			per bit		
		2bit	4bit	8bit	2bit	4bit	8bit
A	No. of 4 input LUTs	16	32	64	8	8	8
	equiv. gate count	96	192	384	48	48	48
B	No. of 4 input LUTs	16	32	64	8	8	8
	equiv. gate count	96	192	384	48	48	48
C	No. of 4 input LUTs	16	32	64	8	8	8
	equiv. gate count	96	192	384	48	48	48

The equivalent gate count per bit is constant within one architecture, as well as for the different architectures. Thus the absolute gate count increases linearly with the width of the signal vector. The equivalent gate count does not include the routing, but this is considered negligible.

From these results it can be concluded that the resource occupation is independent from the SHC architecture. However, the latter might affect the reconfiguration time and the complexity of the reconfiguration unit due to the different number of reconfiguration inputs that need to be controlled.

5.3. Reconfiguration of Self-Healing Cells

Comparison of Fault Tolerance

To compare the fault tolerance of the three different architectures, first the relevant signals were determined. These signals are the SHC inputs a_{nom}, b_{nom}, a_{red}, b_{red}, the internal logic inputs $a1_{int}$, $b1_{int}$, $a2_{int}$, $b2_{int}$, the SHC outputs c_{nom} and c_{red} and – depending on the architecture – up to four reconfiguration inputs $(R_1..R_4)$. The switches and interconnects are covered implicitly since such an erroneous resource would lead to a defect on one of the mentioned signals. The logic circuit implemented in the SHC is not relevant for this analysis and has not been considered.

A correct case is defined as a situation where at least one of the two SHC outputs (nominal or redundant) is correct. All signals can have the status 'correct' or 'defect', where a defective signal is defined as one that is affected by a stuck-at-1 or stuck-at-0 fault. Then all possible combinations were permuted and rules were applied to exclude those combinations, that would lead to an error on both the nominal and redundant output. Such a case is defined as an *SHC failure*. With this approach it is possible to investigate the fault tolerance even for multiple faults.

As an example, Figure 5.10 shows a situation with two defective signals, a_{nom} and $b2_{int}$. In the nominal configuration both the nominal and redundant logic use the nominal input signals. Thus, the depicted situation in Figure 5.10a would lead to an error on both outputs c_{nom} and c_{red}. The fault on a_{nom} can be mitigated by switching reconfiguration input R_1, so that the redundant input signals are used for the nominal path (Figure 5.10b). Then the nominal output will be correct. The redundant output remains incorrect since one of the internal logic inputs is defective, which cannot be repaired.

However, Figure 5.10b assumes that $R1$ can be switched to '1'. This is either possible if $R1$ is not affected by a fault, or if it is defective *but* is affected by a stuck-at-1 fault. Since we assume an equal probability for stuck-at-1 and stuck-at-0 faults, this gives a 50% probability to achieve a working configuration even if the reconfiguration signal is defective.

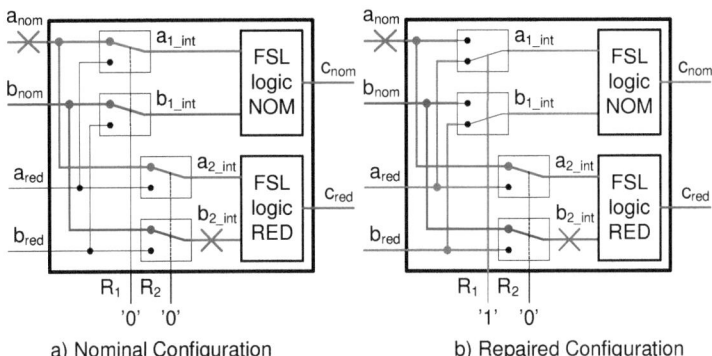

Figure 5.10: SHC Affected by Multiple Faults

For this reason two analyses, one for the worst case and one for the best case, were performed. In the worst case it is assumed that the reconfiguration signal will always have the wrong state if it is defective, whereas in the best case it always has the correct state anyhow, and thus the reconfiguration will be successful. The exclusion rules are as follows:

1. For each SHC input: if the SHC input is defective and (i) either the reconfiguration input or (ii) the corresponding redundant SHC input is defective, then the internal logic input is incorrect. In the worst case (i) and (ii) apply, in the best case only (ii) applies.

2. If the internal logic input is incorrect or defective, the corresponding SHC output is incorrect.

3. If both the nominal and redundant SHC output are incorrect or defective, this fault combination is a 'fail' case. Otherwise the combination is a 'success'.

Figure 5.11 shows the results of this analysis. It presents the number of 'success' cases in relation to the percentage of defective signals for the individual architectures. The absolute number of defective signals varies between the different architectures and between the worst and best case, since in the best case the reconfiguration inputs are considered to have the correct state even if they are defect.

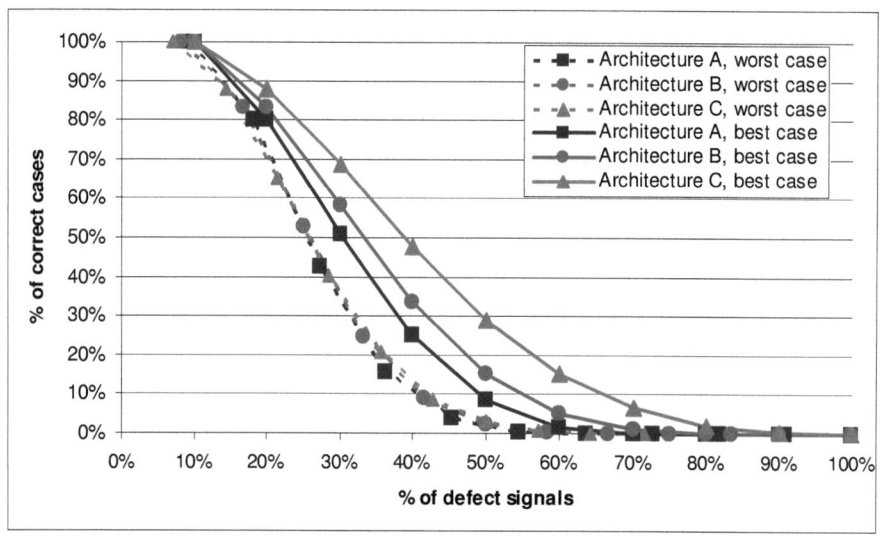

Figure 5.11: Comparison of SHC Architecture vs. Fault Tolerance

For the worst case all three architectures show a similar fault tolerance ('bundle' of curves on the left). For the best case, however, in particular for the range of 20% to 70% of defective signals, architecture C shows a significantly better multiple fault tolerance than architectures A and B. Single faults can be handled with all architectures in both the worst and best case.

5.4 Simulation of Pipeline Reconfiguration

5.4.1 Simulation of a Deadlock Recovery

To illustrate the recovery procedure a first simple implementation of an autonomous deadlock recovery is shown in Figure 5.12. This simulation was performed using Modelsim, with circuits coded in VHDL.

The watchdog counter (in the deadlock detector) is reset by the phase detectors in the FSL registers. If it wraps around, a new configuration is requested by asserting *Req*. The reconfiguration unit comprises a counter that is incremented with each request. After the new reconfiguration has been applied, the *Ack* signal is asserted, which resets the deadlock detector. If the configuration was successful, the circuit's operation continues, preventing any further requests. If not, a new setting will be requested after the deadlock timeout has expired until a working configuration is found.

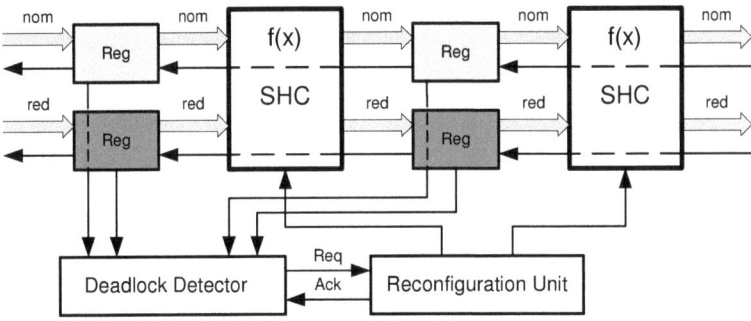

Figure 5.12: Deadlock Detection and Reconfiguration

The logic $f(x)$ contained a fine granular 4-bit ripple carry adder (the principle design is shown in Figure 5.13), where each basic gate was replaced by a SHC. The circuit comprises 34 reconfiguration inputs. All SHCs are initialized to use their nominal input by default. Figure 5.14 shows the simulation of a permanent fault injected into the carry bit calculation of the LSB. More precisely, the a-rail of one input of the nominal AND gate that calculates the carry bit in a half adder is forced to logic 1.

The inputs of gate HA0_AND2_NOM are $a1 = 00$ and $b1 = 11$, which nominally results in an output of $c1 = 00$. Due to the permanent fault, the output is stuck at $c1 = 10$, which generates a wrong phase and produces a deadlock. Examining the reconfiguration inputs shows that bits 8-11 have to be set to logic 1 to select the redundant carry bit. To save time, the simulation starts with a reconfiguration setting of 0x000000EFE. The circuit is halted due to the permanent fault and the deadlock unit starts generating requests for the reconfiguration unit. When the reconfiguration input is set to 0x000000F00, the redundant carry bit is selected, which holds the correct value $c1 = 00$ and the circuit resumes its operation, which is indicated by the activity on the data lines.

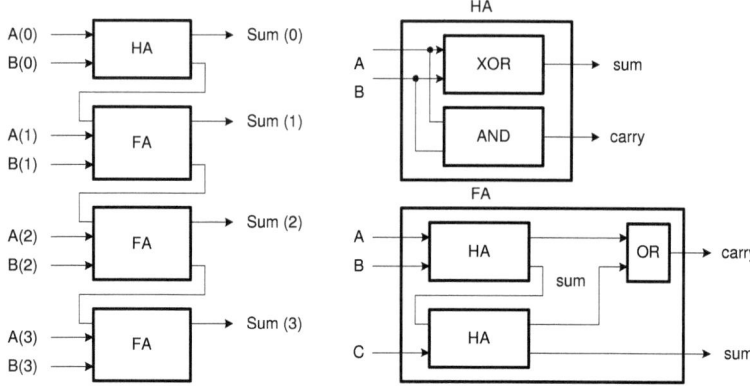

Figure 5.13: Adder Circuit

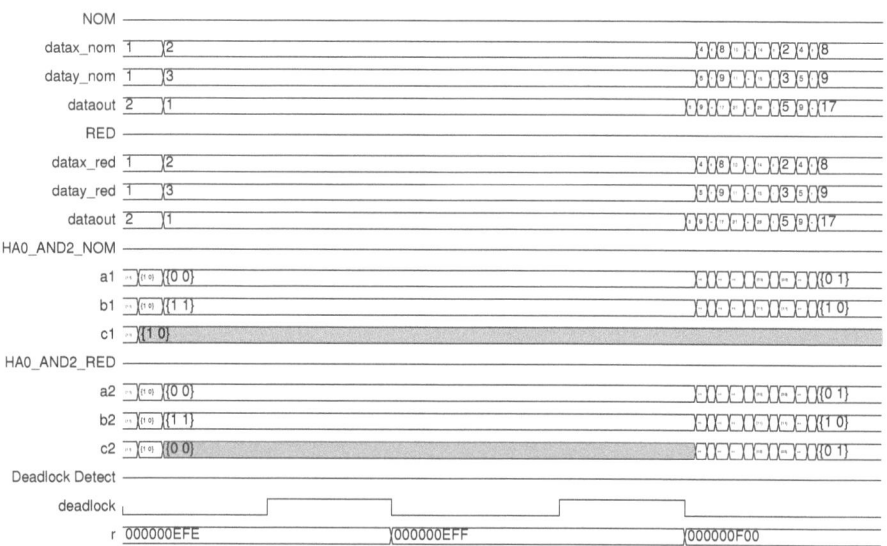

Figure 5.14: Reconfiguration Simulation

5.4.2 Simulation Results of Different Reconfiguration Algorithms

Initially, the idea was to randomly "search" for a configuration that brings the circuit back to operation after a permanent fault has occurred. This would make diagnosis obsolete. However, it turned out that a stochastic approach can cause incorrect data within the pipeline that cannot be detected on this level (details are described later in this section). As a consequence, the pipeline status in case of a fault has to be regarded and particular rules have to be followed, i.e. an algorithm has to be applied. In order to evaluate the effectiveness, efficiency and behavior of different reconfiguration algorithms a simulation model was established (see section 5.2.2).

The following paragraphs describe the results and experiences obtained with this model. Different reconfiguration algorithms were applied to a 5-stage pipeline with up to three injected faults.

While for the evaluation of valid data and acknowledge paths in the pipeline all faults could be injected simultaneously, because the paths are checked without any dependency on previous reconfigurations (section 4.5), a real reconfiguration algorithm only tries some particular routes according to the defined rules. Therefore, the sequence of multiple fault injection can play an important role, as the determination of the fault type, and thus its reconfiguration, could change depending on previous activities.

So, the faults were injected one-at-a-time. After each fault injection the reconfiguration unit checked for a deadlock and executed a reconfiguration, if necessary. If a fault could not be repaired, this sequence was stopped and declared to be "un-repairable". This was repeated for all permutations of multiple faults in order to determine the influence of the fault sequence. This evaluation is depicted in Figure 5.15

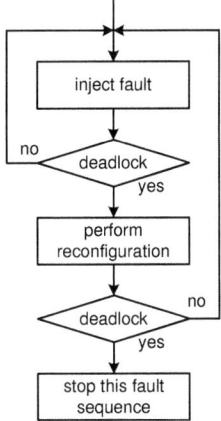

Figure 5.15: Evaluation of Reconfiguration Algorithm

As a general rule, faults at SHC inputs were repaired by changing both the nominal and redundant path of the SHC to use the redundant input data. Basically, faults at nominal SHC inputs could be repaired by changing only the redundant reconfiguration bit of the affected

5. Analysis, Simulations and Experimental Results

SHC. However, as the SHC input will remain defective for a permanent fault, it does not make sense to stay with the nominal configuration for the nominal input.

As will be described below, toggling reconfiguration signals can cause troubles in the pipeline. Therefore any changed reconfiguration bit was never reset again in all reconfiguration units described hereafter.

The designators for the various reconfiguration algorithms are related to identifiers used during the integration of the simulation environment and do not follow any particular nomenclature. Thus, some designators (e.g. A and B) are missing in the subsequent descriptions.

RU_Stochastic

This reconfiguration unit was implemented as simple counter where the counter output controlled the reconfiguration inputs of the application. Each time a deadlock occurred, the counter was increased by one step. With this approach no evaluation of any internal pipeline symptoms would be necessary, as the reconfiguration simply progresses until pipeline activity is detected again. All possible configurations would be tried, and after a wrap around the whole range of configurations is available again.

However, although this approach elegantly exploits the benefits of asynchronous design (see section 2.3), it turned out to be troublesome due to the following reason: For faults at register inputs and acknowledge signals the phase conditions will be different for the nominal and redundant pipeline path after the fault location (see sections 4.4.3 and 4.4.4). If the configuration bits of working elements located later in the pipeline are then toggled, every second time the expected phase will appear and thus cause the data being captured by the register. However, it might not be the correct data but the last data stored in a previous register, which is not updated any more due to the deadlock. Finally, this will cause alternating, consistent but wrong data at the output due to an up-counting reconfiguration controller.

The listing 5.1 below shows an example of such a situation with a fault at the input of Reg5. The nominal path is in a deadlock condition and the redundant output toggles between the value '5' and '6'.

Another disadvantage of the stochastic approach is the potentially large reconfiguration time of $t_{reconfigure} = t_{WD\_timeout} * \sharp reconfigurations$. If an n-bit counter is used the reconfiguration might take up to $2^n * t_{WD\_timeout}$, which will soon reach unacceptable values even for short watchdog timeouts.

RU_C

This reconfiguration unit evaluates the symptoms at the first register of the pipeline only to determine the reconfiguration rules (*global view*, see section 4.4.5), but checks the *Done* and *Pass* signals of all registers to determine the fault location. This method is suitable for a single reconfiguration controller which handles the whole pipeline.

The reconfiguration rules after occurrence of a deadlock are as follows:

1. Check the phase of the pipeline inputs and determine the type of the fault

2. Start from the end of pipeline and go to begin of the pipeline

3. Locate the fault according to the symptoms defined in section 4.4.5

5.4. Simulation of Pipeline Reconfiguration

Listing 5.1: Phase Toggling with Counter RU

```
ConfigVector: 0000000000
src    SHC1   Reg1   SHC2   Reg2   SHC3   Reg3   SHC4   Reg4   SHC5   Reg5
9__9_____9____8_____8____7_____7____6_____6____5____X_4_____ 4
    \          \          \          \          \          \
     \          \          \          \          \          \
      \          \          \          \          \          \
0__0   \__9____9   \__8____8   \__7____7   \__6____6   \__5_____ 5

ConfigVector: 0000000001
src    SHC1   Reg1   SHC2   Reg2   SHC3   Reg3   SHC4   Reg4   SHC5   Reg5
9__9_____9____8_____8____7_____7____6_____6____5____X_4_____ 4
    \          \          \          \
     \          \          \          \
      \          \          \          \
0__0   \__9____9   \__8____8   \__7____7   \__6____6_____6_____ 6

ConfigVector: 0000000010
src    SHC1   Reg1   SHC2   Reg2   SHC3   Reg3   SHC4   Reg4   SHC5   Reg5
9__9_____9____8_____8____7_____7____6_____6____5    X_4_____ 4
    \          \          \          \          \  /
     \          \          \          \          \/
      \          \          \          \          /\
0__0   \__9____9   \__8____8   \__7____7   \__6____6/   \__5_____ 5

ConfigVector: 0000000011
src    SHC1   Reg1   SHC2   Reg2   SHC3   Reg3   SHC4   Reg4   SHC5   Reg5
9__9_____9____8_____8____7_____7____6_____6____5    X_4_____ 4
    \          \          \          \          /
     \          \          \          \        /
      \          \          \          \      /
0__0   \__9____9   \__8____8   \__7____7   \__6____6/_____6_____ 6
```

4. For a fault at a SHC input: reconfigure the SHC located subsequently to the suspect register to use the redundant input for both the nominal and redundant logic

5. For a fault at a register input: split the data path by reconfiguring all SHCs to use the nominal inputs for the nominal logic and the redundant inputs for the redundant logic in order to split the data paths (see section 4.3.2)

6. For a fault at an acknowledge signal: split the data path by reconfiguring all SHCs to use the nominal inputs for the nominal logic and the redundant inputs for the redundant logic

7. If no suspect register has been identified: the fault is located at the input of the first SHC in the pipeline. Reconfigure the first SHC to use the redundant input for both the nominal and redundant logic

This reconfiguration controller works well for single faults and even comes close to the theoretical maximum of reconfigurable conditions for double faults. However, it fails in the following situations with multiple faults: If the first fault blocks the redundant handshake, i.e.

5. Analysis, Simulations and Experimental Results

the fault is located at a redundant register input or at a redundant acknowledge signal, it will not cause an immediate deadlock but only stop the processing of tokens in the redundant path up to the fault location. As separated sources for the nominal and the redundant pipeline path are used, only the redundant source will not provide any new tokens to the pipeline. A second fault, occurring in the nominal path, will cause a deadlock. The recovery actions of this reconfiguration unit are based on the phase information of the pipeline input. If the second fault is located later in the pipeline than the first fault, the phase of the redundant pipeline input depends on the number of tokens that have been processed between the two fault occurrences. Thus, there is a 50% probability that the correct recovery action is selected.

The listings 5.2 and 5.3 show the same double-fault situation but with one more token processed between the two fault occurrences in the second case. The first situation fails because the symptoms lead to an acknowledge fault where the pipeline paths become separated. Since there is another fault in the nominal path, this reconfiguration does not work. In the second case the SHC fault is observed, which is reconfigured locally and leads to a working pipeline configuration.

Listing 5.2: Example where RU_C fails
```
 1 Deadlock condition after second fault:
 2 src   SHC1   Reg1   SHC2   Reg2   SHC3   Reg3   SHC4   Reg4   SHC5   Reg5
 3 7__7_____7____6_____6____5_____5____4_____4____X_____2_____ 2
 4      \          \          \          \          \
 5       \          \          \          \          \
 6        \          \          \          \          \
 7 6__6   \__7____5   \__6____4   \__5____4   \__4____3   \__2_____ 2
 8                    X
 9 Phases:
10 src   SHC1   Reg1   SHC2   Reg2   SHC3   Reg3   SHC4   Reg4   SHC5   Reg5
11 0___1_____1__0_____0__1_____1__0_____0__1_____0__0__
12      \      0   1  \      1   0  \      0   1  \      1   0  \      0   0
13       \              \              \              \              \
14        \              \              \              \              \
15 1___0   \__1__1__   \__0__0__   \__1__0__   \__0__1__   \__0__0__
16          1   0          0   X          0   1          1   0          0   0
17
18 Recovery condition (ConfigVector: 0101010101):
19 src   SHC1   Reg1   SHC2   Reg2   SHC3   Reg3   SHC4   Reg4   SHC5   Reg5
20 7__7_____7____6_____6____5_____5____4_____4____X_____2_____ 2
21
22
23
24 6__6_____7____5_____6____4_____5____4_____4____3_____2_____ 2
25                    X
```

5.4. Simulation of Pipeline Reconfiguration

Listing 5.3: Example where RU_C succeeds in same condition
```
 1 Deadlock condition after second fault:
 2 src  SHC1    Reg1  SHC2    Reg2  SHC3    Reg3  SHC4    Reg4  SHC5    Reg5
 3 8__8_____8____7_____7____6_____6____5_____5____X_____3_____ 3
 4      \           \           \           \           \
 5       \           \           \           \           \
 6        \           \           \           \           \
 7 6__6    \__8____5   \__7____4   \__6____5   \__5____4   \__3_____ 3
 8                          X
 9 Phases:
10 src  SHC1    Reg1  SHC2    Reg2  SHC3    Reg3  SHC4    Reg4  SHC5    Reg5
11 1___0_____0__1_____1__0_____0__1_____1__0_____1__1__
12      \     1  0  \     0  1  \     1  0  \     0  1  \     1  1
13       \           \           \           \           \
14        \           \           \           \           \
15 1___0    \__0__1__   \__1__0__   \__0__1__   \__1__0__   \__1__1__
16           1  0        0  X        1  0        0  1        1  1
17
18 Recovery condition (ConfigVector: 0000000011):
19 src  SHC1    Reg1  SHC2    Reg2  SHC3    Reg3  SHC4    Reg4  SHC5    Reg5
20 8__8_____8____7_____7____6_____6____5_____5____X     __3_____ 3
21      \           \           \           \           /
22       \           \           \           \         /
23        \           \           \           \       /
24 6__6    \__8____5   \__7____4   \__6____5   \__5____4/_____3_____ 3
25                          X
```

RU_D

This reconfiguration controller considers the local symptoms presented in section 4.4.5.
The reconfiguration rules after occurrence of a deadlock are as follows:

1. Start from the end of the pipeline and go to the begin of the pipeline

2. Locate the fault according to the symptoms defined in section 4.4.5

3. For a fault at a SHC input: reconfigure the SHC located subsequently to the suspect register to use the redundant input for both the nominal and redundant logic

4. For a fault at a REG input: reconfigure all SHCs up to the suspect register to use the redundant input for the redundant output + configure the subsequent SHC to use the redundant input for both the nominal and redundant output (section 4.3.3).

5. For a fault at an ACK signal: reconfigure all SHCs up to the suspect register to use the redundant input for the redundant output + configure the subsequent SHC to use the redundant input for both the nominal and redundant output (section 4.3.4)

6. For a fault at the input of the last register: reconfigure all SHCs to use the redundant input for the redundant output. This needs to be done because the nominal path is blocked and there is no possibility to bypass the nominal register, as there are no reconfigurable elements after the last register

7. For a fault at the acknowledge signal of the last register: reconfigure all SHCs to use the redundant input for the redundant output

8. If no suspicious register has been identified: the fault is located at the input of the first SHC in the pipeline. Reconfigure the first SHC to use the redundant input for both the nominal and redundant logic

9. If the SHC preceding or following the suspect register has been reconfigured previously, the symptoms would lead to a fault at a SHC input, but it is a fault between the suspect and the next register. In this case all SHCs shall be reconfigured to use the redundant input for the redundant output. This rule is an exception which provides an improvement of the fault detection.

This reconfiguration unit is well suited for a distributed reconfiguration controller, as it moves from one register to the next and takes the decisions based on the local phase conditions.

5.4. Simulation of Pipeline Reconfiguration

RU_E

This reconfiguration unit considers the local symptoms presented in section 4.4.5 and also uses acknowledge switches for the reconfiguration. The default configuration of the acknowledge switches was chosen in a way that the nominal pipeline receives the nominal acknowledge and the redundant pipeline receives the redundant acknowledge. RU_E is also suitable for a distributed reconfiguration unit.

The reconfiguration rules are as follows:

1. Start from end of pipeline and go to begin of the pipeline

2. Locate the fault according to the symptoms mentioned in section 4.4.5

3. For a fault at a SHC input: reconfigure the SHC subsequent to the suspect register to use the redundant input for both the nominal and redundant logic

4. For a fault at a REG input: reconfigure the SHC subsequent to the suspect register to use the redundant input for both the nominal and redundant output (section 4.3.3) and the acknowledge switch between the suspect and the preceding register to use the redundant acknowledge for the nominal pipeline path.

5. For a fault at an ACK signal: reconfigure the SHC subsequent to the suspect register to use the redundant input for both the nominal and redundant output (section 4.3.4) and the acknowledge switch between the suspect and the preceding register to use the redundant acknowledge for the nominal pipeline path

6. For a fault at the input of the last register: reconfigure the last acknowledge switch to use the redundant acknowledge for the nominal pipeline path

7. For a fault at the acknowledge of the last register: reconfigure the last acknowledge switch to use the redundant acknowledge for the nominal pipeline path

8. If no suspicious register has been identified: the fault is located at the input of the first SHC in the pipeline. Reconfigure the first SHC to use the redundant input for both the nominal and redundant logic

The simulations of this reconfiguration unit revealed the following problems:

1. Tokens might be lost and cause a subsequent uncorrectable deadlock although a valid data and acknowledge path has been established. This phenomenon can occur if a nominal acknowledge signal is affected by a permanent fault and a previous fault at a SHC input earlier in the pipeline caused the data path to be separated from the acknowledge path. Listing 5.4 shows the final status of the pipeline in such a condition (the first fault was at the input of SHC2, the second one at the *Pass* signal of REG4). The handshake of the nominal acknowledge path will be blocked due to the second fault and the registers will be filled up with new tokens up to the affected register. Since the redundant path is not affected by a fault, it will capture one more token than the nominal path. However, since data and handshake path are separated between the nominal and the redundant path due to the first fault at the SHC input, the redundant path defines the input of both the nominal and the redundant register located subsequently to the reconfigured

SHC. Thus, although the nominal register has not updated its *Done* signal, it receives new input data (the one-after-the-next token) and looses the token in between (input vs. output of Reg2). Since the redundant path captured one more token than the nominal path, the input and the output of the nominal register in this stage then have the same phase. Consequently, the condition that the output phase must be different to the input phase to become transparent cannot be achieved any more, and the pipeline will end up with a permanent and uncorrectable deadlock.

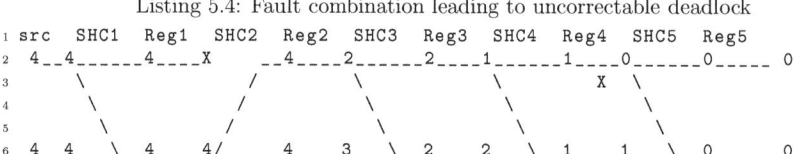

Listing 5.4: Fault combination leading to uncorrectable deadlock

```
1  src     SHC1    Reg1    SHC2    Reg2    SHC3    Reg3    SHC4    Reg4    SHC5    Reg5
2  4__4_____4____X    __4____2_____2____1_____1____0_____0_____ 0
3          \                   /           \               \           X   \
4           \                 /             \               \               \
5            \               /               \               \               \
6  4__4   \__4____4/_____4____3   \__2____2   \__1____1   \__0_____ 0
```

2. In the initial default configuration, a fault leading to a deadlock must have occurred in the nominal path. After a reconfiguration has been performed, this might not be true any more. With the observation possibilities described so far, a fault at a nominal signal (acknowledge signal or register input) cannot be distinguished from a fault at a redundant signal. However, the reconfiguration must be performed differently for nominal and redundant faults. Thus, after a reconfiguration the fault location is not known accurately enough to decide which configuration to be used. This means that the chosen configuration will fail with 50% probability.

3. Since a fault at a register input or acknowledge signal will cause the register to not update the output any more even after a successful reconfiguration, this might cause the following troubles: If a redundant register input or acknowledge signal is affected by the first fault, this will not stop the pipeline. A subsequent nominal fault earlier in the pipeline will cause a deadlock and trigger the reconfiguration controller to find out the fault location. However, depending on the number of processed tokens between the occurrence of the two faults, the *Pass* and *Done* of the redundant register can be different and thus the reconfiguration unit might find the fault in the redundant path first. Since the difference between *Pass* and *Done* will never disappear, the fault earlier in the pipeline will never be found. The same is true if the first fault at a nominal register input or acknowledge signal was successfully reconfigured and the subsequent fault occurs earlier in the pipeline. This requires more intelligent reconfiguration units, which consider previous configurations.

4. Due to the split of the acknowledge and data path and its independent routing some configurations can lead to timing problems (the data path is timed by the nominal pipeline, the acknowledge path by the redundant pipeline). This was not observed in this simulation, but in the hardware implementation described in section 5.5.

5.4. Simulation of Pipeline Reconfiguration

RU_F

This reconfiguration controller is based on RU_E, but also considers previous reconfigurations. From the results with RU_E it was obtained that the acknowledge path should follow the data path. Thus, the default configuration of the acknowledge switches was changed in a way, that both the nominal and redundant pipeline receive the nominal acknowledge.

The following rules were added:

1. If the SHC located subsequently to the suspect register (if applicable, i.e. not for the last register in the pipeline) and the acknowledge switch preceding the suspect register was previously reconfigured (configuration is different to default configuration), skip this cycle and move on with checking for symptoms at the next (preceding) register.

2. For reconfigurations due to faults at register inputs or acknowledge signals: reconfigure the acknowledge switches surrounding the SHC being reconfigured in a way, so that the acknowledge path follows the data path.

These rules need to be applied in addition to the already defined rules. For example, the fault location (suspect register) needs to be identified as described for RU_E, but then it must be checked if the subsequent SHC and/or acknowledge switch has been reconfigured previously. If this is the case, the additional rule 1) applies (i.e. no reconfiguration), otherwise the reconfiguration as defined for RU_E shall be performed.

Three variants were simulated, with only one of the above additional rules and both rules implemented, in order to see the influence of the distinct improvements.

For the final variant (both rules added), the following improvements in the results could be observed compared to RU_E :

- Problem 1) of RU_E is solved due to the changed default configuration and the additional rule, that the acknowledge path shall follow the data path.

- Problem 2) of RU_E remains but is only relevant for multiple faults in the pipeline.

- Problem 3) is solved because earlier reconfigurations are considered in the reconfiguration rules. If a situation *"Pass* is unequal to *Done"* at a register is detected but either the subsequent SHC or the preceding AS has been reconfigured previously, this stage is skipped in the reconfiguration sequence and the next (preceding) register is checked.

- Problem 4) is mitigated: due to the rule that the acknowledge path follows the data path, the circuit timings are controlled in a better way.

5.4.3 Result Summary and Comparison

The following table 5.4 summarizes the simulation results of the different reconfiguration units for up to three faults. Analysis with a higher number of faults could not be performed due to the combinational explosion of possibilities and the resulting high simulation time. Furthermore, as can be seen in the table, the probability that three faults in the pipeline can be handled, is less than 60% in the best case. With more than three faults the probability is expected to drop below 50%, which is thus not really useful any more from a practical point of view.

For each reconfiguration unit the absolute and relative value of successfully handled fault situations (not necessarily repairs) are given. Since all possible sequences (permutations) of the fault injection were applied, the number of fault combinations presented previously in Table 4.8 is multiplied by $n!$ with n being the number of faults. Although there is no dependency on the sequence for the theoretical values due to the way they are determined (see section 4.5), the theoretical values from table 4.8 have been multiplied as well for comparison.

RU_C comes very close to the theoretical number of possible reconfigurations. A dependency of the success rate was observed for different fault sequences (0.5% difference for double faults, 0.2% for triple faults), the reasons are described above. In table 5.4 the lowest number (percentage of successful repairs) is shown.

RU_D4 does not show any dependence on the fault sequence and can repair all double faults that have been identified to be repairable. For triple faults nearly the theoretical number is reached.

RU_E uses acknowledge switches but splits the data and acknowledge path. For double faults this reconfiguration unit provides better results than the units without using acknowledge switches, but for triple faults the performance decreases significantly and is lower than that from RU_C and RU_D4 without acknowledge switches. The reason is that a lot of faults are incorrectly identified and thus the selected reconfiguration is not appropriate.

RU_F is based on RU_E, but with the difference that the acknowledge path follows the data path. The results improve significantly, in particular for triple faults.

RU_F1 is also based on RU_E, but considers previous configurations and skips cycles which could lead to wrong reconfigurations. The results also improve significantly compared to RU_E.

RU_F3 considers previous configurations and takes care that the acknowledge path follows the data path. This reconfiguration unit provides the best results of all simulated algorithms.

Some of the remaining situations could be solved by very specific measures and explicit handling, so that the fault tolerance for double faults could be improved to 85.6%. However, as no general rule could be found, this approach is not considered feasible and not listed in the table. Overall, the variants of RU_F achieve significantly better results than the algorithms without using acknowledge switches.

It shall be noted that the results consider pipeline faults only, i.e. faults at SHC inputs, register inputs and at acknowledge signals, but do not cover faults in (complex) SHCs. Due to the high fault tolerance of SHCs the overall improvement will be even higher. Refer to section 5.8 for more information.

Based on the described symptoms it is not possible to distinguish between faults in the nominal or redundant path (see the explanation for RU_E). The rules could be improved by more intelligent reconfiguration units, which check more details of the pipeline and consider differences between faults in the two paths. However, this has not been done for the presented reconfiguration units.

5.4. Simulation of Pipeline Reconfiguration

Although the self-healing architecture provides quite a lot of reconfiguration possibilities (refer to section 4.3), only a subset of them is needed and actually used in the reconfiguration units. This gives the opportunity to either simplify the design of the configurable elements (e.g. architecture A of the SHCs could be used for coarse granular implementations, refer to section 5.3.2), or to make use of the remaining configurations to increase the fault tolerance. Such optimizations of the reconfiguration units are, however, out of scope of this thesis and have not been further treated.

As described in section 4.4.5, in most cases a successful repair with a pipeline structure using acknowledge switches re-establishes a situation like in the fault-free configuration, and causes the same symptoms for subsequent faults. The rare cases where the application of the "single fault diagnosis" to multiple faults leads to a wrong identification would need to be handled by a rather intelligent reconfiguration unit. However, as this is only a problem in particular fault combinations (e.g. the faults occur in neighboring elements) and only affects the multiple fault tolerance, it was considered acceptable to use the same reconfiguration rules as for single faults also for subsequent faults. This also reduces the complexity of the reconfiguration unit.

The results justify the usefulness of the approach: A probability of $> 80\%$ for tolerating double faults and $\sim 60\%$ for triple faults is significantly better than a TMR system (see section 5.8).

without acknowledge switches		theoretical		RU C		RU D4	
faults	total	abs.	pct.	abs.	pct.	abs.	pct.
1	30	30	100.0%	30	100.0%	30	100.0%
2	870	508	58.4%	480	54.7%	508	58.4%
3	24360	6756	27.7%	6190	25.4%	6594	27.1%

with acknowledge switches		theoretical		RU E		RU F		RU F1		RU F3	
faults	total	abs.	pct.	abs.	pct.	abs.	pct.	abs.	pct.	abs.	pct.
1	30	30	100.0%	30	100.0%	30	100.0%	30	100.0%	30	100.0%
2	870	788	90.6%	565	64.9%	690	79.3%	617	70.9%	725	83.3%
3	24360	17928	73.6%	2760	11.2%	12702	52.1%	11566	47.5%	14322	58.8%

Table 5.4: Comparison of Reconfiguration Algorithms (Simulation Results)

5.5 Hardware Fault Injection Experiments

5.5.1 General

The hardware fault injection experiments aimed to prove the concept of the self-healing architecture, the correctness of the simulations, the validity of the simplification made, and should highlight any effects that were not observed in the simulations, e.g. due to timing problems in the circuit. For a detailed description of the used environment refer to section 5.2.3.

Two circuits were implemented in an FPGA:

1. A 5-stage pipeline without acknowledge switches and using the reconfiguration unit RU-D4

2. A 5-stage pipeline with acknowledge switches and using the reconfiguration unit RU-F3

The reconfiguration units, which are basically simple state machines, were implemented as synchronous circuits to save resources and to avoid any side-effects. A clock frequency of 100MHz was used, and the watchdog timeout was chosen to be 500 clock cycles, i.e. $5\mu s$. The reconfiguration takes 3 clock ticks to start, 2 clock ticks per pipeline stage (the number of reconfiguration controllers equals the number of pipeline stages in a distributed approach, such as with RU_F3), and 1 clock tick to set the reconfiguration pattern. This results in a total maximum reconfiguration time of 14 clock cycles for the implemented circuits, equal to 140ns.

The target device for the hardware experiments was a Xilinx Virtex-4 FPGA. Although the author is aware that using this device might not lead to resource-efficient and fully representative implementations of asynchronous circuits, this approach is considered appropriate to prove the concept in hardware.

The circuit is shown in Figure 5.16. The small rectangles designate saboteurs which allow injecting faults at particular signals. At FSL vectors rail 'a' of bit '0' was taken as representative and subjected to fault injections to generate a token fault.

The same situations as simulated with the Matlab model (see section 5.4) were applied to the circuit and compared to the simulation results. In contrast to the Matlab simulation, where the faulty state of an FSL vector was considered in an abstract way (vector is either valid or inconsistent), the faults for the hardware fault injection experiments were injected on a single rail of a 4-bit FSL vector. This implies that – depending on the phase of the vector at the time the fault is injected – the fault can either have an immediate effect if it changes the state of the affected rail, or it is latent for one or more asynchronous processing cycles, if the rail currently has the same state as the applied stuck-at fault. The fault will not lead to a deadlock until it causes an inconsistent vector.

To highlight this behavior, four data sequences as shown below were applied. Each sequence consists of 20 values with different distributions of ones and zeros. Furthermore, when multiple faults were injected, the faults were applied at different phases. Each fault injection scenario was run twice to identify potential timing problems.

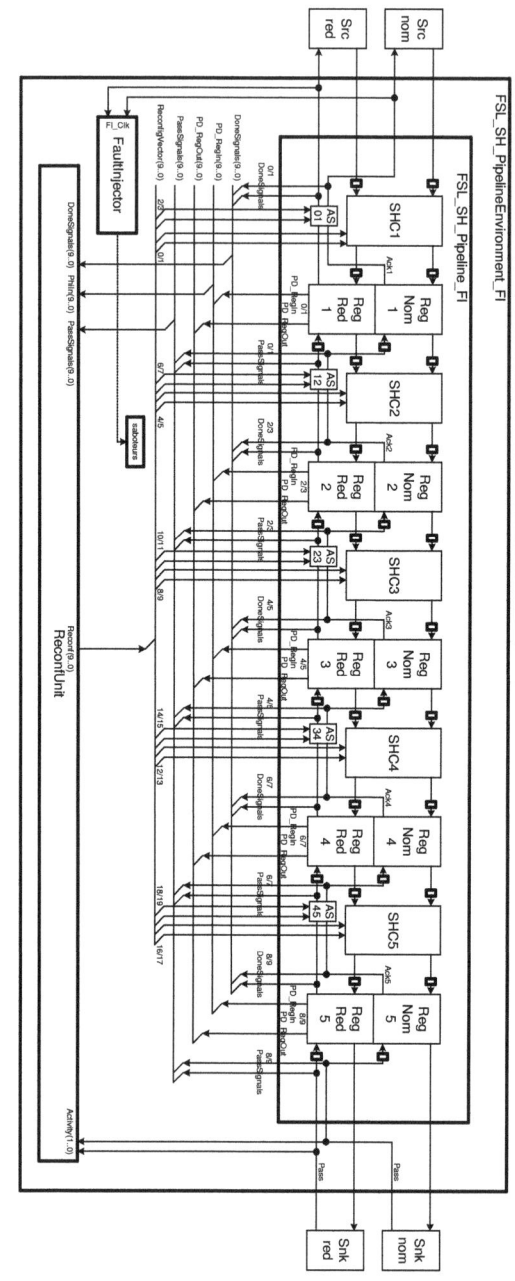

Figure 5.16: Circuit used for Hardware Fault Injection Experiments with Pipeline Reconfiguration

5.5. Hardware Fault Injection Experiments

```
Data sequences (values in hexadecimal):

1: 1 2 3 4 5 6 7 8 9 A B C D E F 1 2 3 4 5
2: 5 4 3 2 1 F E D C B A 9 8 7 6 5 4 3 2 1
3: 5 A 5 A 5 A 5 A 5 A 5 A 5 A 5 A 5 A 5 A
4: F 1 E 2 D 3 C 4 B 5 A 6 9 7 8 F 7 E 6 D
```

The SHCs did not contain any logic function but only switches to route the input signal through to the output. This makes the model insensitive to effects caused by hazards as described in section 2.4.5.

5.5.2 Results of Pipeline without Acknowledge Switches

All single faults could be repaired and worked with all data patterns. With two faults 17 out of 870 fault combinations could be repaired in the simulation but not in hardware. Table 5.5 lists the cases with different results of both experiment runs for all four data sequences. A blank cell indicates that the same result as in the simulation was achieved, an "X" indicates a difference, i.e. this case was repaired successfully in the simulation using reconfiguration unit RU_D4, but not in the hardware experiment. No cases were observed, where the hardware could solve a failure case which failed in the simulation.

It can be seen that for pattern P2 and P4 no differences were observed, while for P1 and P3 17 and 16 differences occurred. Except of four combinations, the nominal input of SHC3 was always affected by a fault. For all unsuccessful combinations the reconfiguration leads to a configuration, where SHC3 uses the redundant input as source for both outputs. This means that the data and acknowledge path are split, i.e. the redundant acknowledge defines the timing of the data in the nominal pipeline path. SHC3 is in the middle of the pipeline and thus has the maximum distance to both the source and the sink, which could lead to an awkward signal routing in the FPGA. Similar fault situations and configurations exist with the other SHCs in the pipeline for which such troubles were not observed.

In listing 5.5 two examples are given, which show that with such a SHC-3 configuration the redundant register Reg-2 becomes the data source, while the acknowledges are separated. Based on the constraints given in section 4.3.5 a violation of the timings can occur if the difference between nominal and redundant path is above approximately 30% in such a configuration. From the place&route report of the FPGA synthesis some timings of the control signals in the FSL registers were extracted (internal phase detector outputs *SetIn* and *SetOut*, control signals of data and handshake latches *LatchEnable* and *DoneEnable*). A comparison of these timings revealed variations between the nominal and redundant path of the same stage of up to approximately 90% (Figure 5.17). Between subsequent stages even differences of more than 120% have been found. The large differences are concentrated on the stages three and four, i.e. the timing conditions are most severe in this area. This corresponds to the observation of the fault injection experiments explained above, where the problems were observed with SHC3 being configured to use the redundant input. In the hardware implementation the SHCs did not contain any sophisticated logic, so the propagation delay t_L was rather short, which makes the timing conditions more severe. Violations of the timing constraints are obviously the reason for the unsuccessful repairs in this particular pipeline configuration. Although the available timing results confirm the theoretical predictions quite well, more details need to be

first fault	second fault	P1		P2		P3		P4	
Reg1.PassRed	SHC3.ValInNom	X	X			X	X		
SHC3.ValInNom	Reg1.PassRed	X	X			X	X		
SHC3.ValInNom	Reg1.PassNom	X	X						
SHC5.ValInRed	SHC3.ValInNom	X	X			X	X		
SHC3.ValInNom	SHC5.ValInRed	X	X			X	X		
SHC4.ValInRed	SHC3.ValInNom	X	X			X	X		
SHC3.ValInNom	SHC4.ValInRed	X	X			X	X		
SHC2.ValInRed	SHC3.ValInNom	X	X			X	X		
SHC3.ValInNom	SHC2.ValInRed	X	X			X	X		
SHC1.ValInRed	SHC3.ValInNom	X	X			X	X		
SHC3.ValInNom	SHC1.ValInRed	X	X			X	X		
SHC3.ValInNom	SHC2.ValInNom	X	X						
SHC3.ValInNom	SHC1.ValInNom	X	X						
SHC3.ValInNom	Reg1.ValInRed	X	X			X	X		
Reg1.ValInRed	SHC3.ValInNom	X	X			X	X		
SHC5.ValInRed	Reg2.ValInNom					X	X		
Reg2.ValInNom	SHC5.ValInRed					X	X		
SHC4.ValInRed	Reg2.ValInNom					X	X		
Reg2.ValInNom	SHC4.ValInRed					X	X		
SHC3.ValInNom	Reg2.ValInNom	X	X						
SHC3.ValInNom	Reg1.ValInNom	X	X						
Differences		17	17	0	0	16	16	0	0

Table 5.5: Cases with Differences between Hardware Experiments and Simulation for the Circuit without Acknowledge Switches using RU_D4

investigated to gain additional confidence. Such investigations were not performed within the scope of this thesis due to insufficient tool support.

Furthermore, there is a dependency on the data pattern: if the fault does not change the signal state (i.e. the signal has the same state as the fault would generate), it does not lead to an immediate deadlock in the next cycle, but one or more tokens could be processed before the fault leads to a deadlock. Since the order and sequence of "high" and "low" states of the individual rails is different for the four patterns, a fault applied at the same time on the same signal (rail) can cause different behavior. Since the tokens are provided to the pipeline in alternating phases, it depends on the number of tokens that are processed and the value of the tokens which phases are visible when the deadlock occurs - and thus wrong symptoms can be identified. Two examples for such situations are described in detail in the next section for the pipeline with acknowledge switches. Similar situations occurred here, which explains the differences between the patterns.

5.5. Hardware Fault Injection Experiments

Listing 5.5: Unsuccessful Reconfiguration without AS in Hardware
```
 1 Faults at SHC3.ValInNom and SHC4.ValInRed:
 2 src   SHC1   Reg1   SHC2   Reg2   SHC3   Reg3   SHC4   Reg4   SHC5   Reg5
 3 8__8_____8____8_____8____X    __8____8_____8____8_____8_____ 8
 4    \          \          /          \          \
 5     \          \        /            \          \
 6      \          \      /              \          \
 7 8__8   \__8____8  \__8____8/_____8____X   \__8____8  \__8_____ 8
 8
 9 Faults at SHC5.ValInRed and Reg2.ValInNom:
10 src   SHC1   Reg1   SHC2   Reg2   SHC3   Reg3   SHC4   Reg4   SHC5   Reg5
11 4__4_____4____3_____X____2    __8____8_____8____8_____8_____ 8
12                             /          \          \
13                            /            \          \
14                           /              \          \
15 8__8_____8____8_____8____8/_____8____8   \__8____X   \__8_____ 8
```

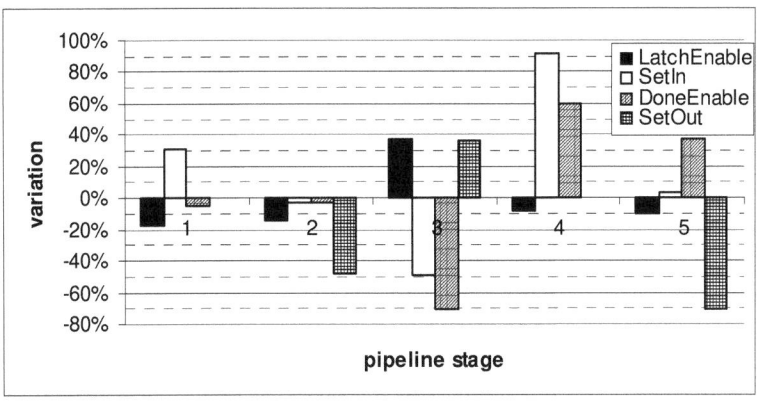

Figure 5.17: Timing Comparison of Nominal vs. Redundant FSL-Register Control Signals

107

5.5.3 Results of Pipeline with Acknowledge Switches

Again, all single faults could be repaired and worked with all data patterns. With two faults 7 differences out of 870 fault combinations were observed between the simulation and the hardware experiments. Table 5.6 lists the cases with different results of both experiment runs for all four data sequences. As before, an "X" designates a difference compared to the simulation, a blank cell indicates that the same result was achieved. All combinations where a difference was found were successfully repaired in the simulation using reconfiguration unit RU_F3.

first fault	second fault	P1		P2		P3		P4	
Reg4.ValInRed	Reg2.PassNom	X	X			X	X		
Reg4.ValInRed	Reg1.PassNom	X	X			X	X		
Reg3.ValInRed	Reg1.PassNom	X	X			X	X		
Reg5.ValInNom	Reg4.PassNom	X	X			X	X		
Reg4.ValInNom	Reg3.PassNom	X	X			X	X		
Reg3.ValInNom	Reg2.PassNom	X	X			X	X		
Reg2.ValInNom	Reg1.PassNom	X	X			X	X		
	Differences	7	7	0	0	7	7	0	0

Table 5.6: Cases with Differences between Hardware Experiments and Simulation for the Circuit with Acknowledge Switches using RU_F3

All differences are related to a second fault on a nominal acknowledge signal and have only been observed in particular cases, which can be distinguished into two scenarios:

Fault at nominal register input (Situation 1) The first fault affects a nominal register input, which will cause a deadlock and initiate a repair. The defined reconfiguration in RU_F3 for this fault is to switch the preceding and the subsequent acknowledge switch and the subsequent SHC to the redundant configuration in order to bypass the defective register (except for the last register in the pipeline, where only the preceding acknowledge switch is changed).

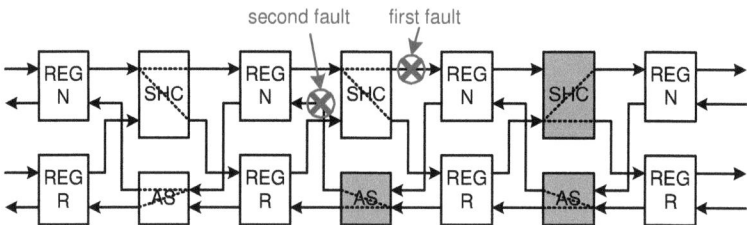

Figure 5.18: Example for Situation 1

A second fault affects the nominal acknowledge signal between the previously affected register and the preceding one, and at a time, where the signal has the same state as the stuck-at fault would generate, i.e. the fault is masked (Figure 5.18 shows an example for such a situation after the first fault has been repaired). The pipeline thus will process tokens until the

5.5. Hardware Fault Injection Experiments

fault causes an invalid data vector leading to a deadlock condition. Depending on the number of tokens processed in the meantime, the registers will have stored tokens in either equal or different phase, so that either the condition for symptom 8 or symptom 9 is fulfilled. The latter one does not meet the expectations of the pipeline behavior in this fault condition and thus will lead to a wrong reconfiguration.

Fault at redundant register input (Situation 2) The first fault is located in the redundant path. It will not lead to a deadlock and remain as latent fault until a subsequent fault affects a nominal signal.

A second fault affects a nominal acknowledge signal earlier in the pipeline at a time, where the signal has the same state as the stuck-at fault would generate, i.e. the fault is masked (Figure 5.19 shows an example for such a situation). The pipeline thus will process tokens until the fault causes an invalid data vector leading to a deadlock condition. Depending on the number of tokens processed in the meantime, either the condition for symptom 2 for the faulty nominal acknowledge signal or symptom 3 for the faulty redundant register is fulfilled. The latter one correctly detects the fault at the redundant register, but as the RU cannot distinguish between faults in the nominal and redundant path, a nominal fault is identified and the defined reconfiguration is applied – which is wrong in this case.

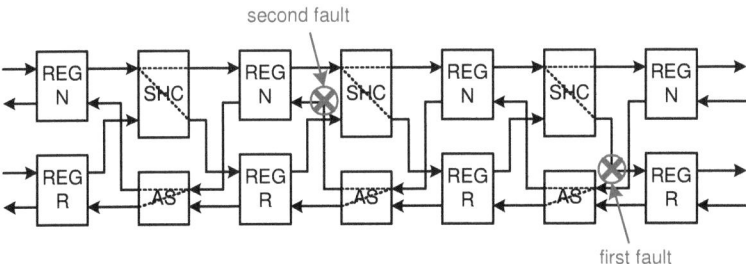

Figure 5.19: Example for Situation 2

5.5.4 Summary

Table 5.7 summarizes the results obtained from the hardware experiments and compares them with the simulation results.

No. of faults	total sequ.	RU D4 sim	RU D4 H/W	RU F3 sim	RU F3 H/W
1	30	30 (100.0%)	30 (100%)	30 (100%)	30 (100%)
2	870	508 (58.4%)	487 (56.0%)	725 (83.3%)	718 (82.5%)
3	24360	6594 (27.1%)	6430 (26.4%)	14322 (58.8%)	13814 (56.7%)

Table 5.7: Result Comparison of Simulation and Hardware Experiments

Faults at SHC inputs can be completely recovered by the SHC reconfiguration, because the defective input is then not used any more. Faults at register inputs remain in the circuit as the input cannot be "switched off".

All single faults could be repaired with the implemented reconfiguration units. For double faults the hardware experiments nearly matched the simulation results. The differences occur due to timing and masking effects, which were not visible in the abstract simulation. For triple faults these effects seem to become even more severe, since the difference between the simulation and the hardware experiments increases in particular for the pipeline with acknowledge switches. Nevertheless, it was proven that the concept can be implemented in hardware and provides the expected results.

5.6 Hardware Implementation of a Complex Self-Healing Circuit

5.6.1 Introduction

In order to prove the suitability of the self-healing approach for complex circuits, the concept was applied to a video processing algorithm as specified for the space mission GAIA.

The *Global Astrometric Interferometer for Astrophysics* (GAIA) is a scientific mission of the *European Space Agency* (ESA) that is scheduled for launch in 2012 [23]. The mission places a large telescope at the Lissajous-type orbit around L2 to generate a precise three-dimensional map of our Galaxy. The *Video Processing Unit* (VPU) provides one of the central functions in GAIA. It pre-processes the digital data acquired by a large CCD array before it will be transmitted to Earth for the final analysis. The algorithms in the GAIA VPU comprise one of the most demanding applications for today's space borne signal processors.

5.6.2 The GAIA Pre-Processing Algorithm

For performance reasons and due to the lack of an adequate, powerful space-compatible processor, the various tasks in the GAIA VPU have been divided into hardware and software based algorithms. For the experiments a portion of the hardware based algorithms was chosen and made self-healing. The following information is retrieved from the official invitation to tender for the VPU and is provided by courtesy of ESA and the prime contractor EADS Astrium [20].

A block diagram of the hardware algorithms in the GAIA VPU is shown in Figure 5.20. For the test application the preprocessing of the star mapper samples was selected, which is encircled in the figure. Actually, these samples are used to identify the stars within the huge amount of data provided by the CCD.

A more detailed view of the star mapper preprocessing is shown in Figure 5.21. Although it only comprises a small portion of the complete VPU algorithms, the preprocessing already includes all typical functions used in signal processing applications, such as saturation checks, multiplication, addition, feedback filters, etc.

5.6. Hardware Implementation of a Complex Self-Healing Circuit

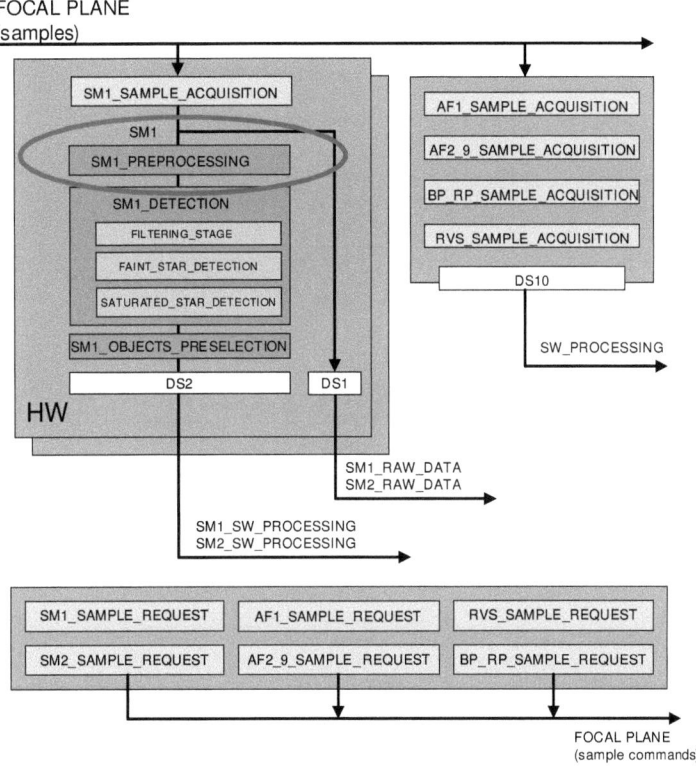

Figure 5.20: GAIA VPU Hardware Algorithms

Actually, it is composed of two main functions:

1. A linear correction checks for saturated values and applies column response and dark signal non-uniformity correction.

2. A dead column correction performs a simple neighbor interpolation for samples coming from pixels that are marked as dead.

The linear correction takes the raw samples $UNPREPRO\_DATA[ac]$ and compares them with a saturation level $SATURATION\_LUT[ac]$ for each CCD row ac. If the sample is saturated, it is replaced by the constant $SATURATED$. Otherwise a linear function correcting the *Dark Signal Non Uniformity* (DSNU) and the *CCD Column Response Non Uniformity* (CRNU) (both are parameters of the CCD) is applied:

5. Analysis, Simulations and Experimental Results

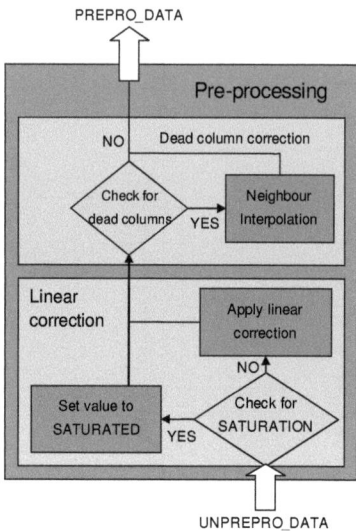

Figure 5.21: Principle of Hardware Pre-Processing

```
1  if UNPREPRO_DATA[ac] > SATURATION_LUT[ac] then
2      PREPRO_DATA[ac] = SATURATED
3  else
4      crnu[ac] = trunc(UNPREPRO_DATA[ac] * CRNU_LUT[ac],15)
5      dsnu[ac] = sat(crnu[ac],16) + DSNU_LUT[ac],16
6
7      PREPRO_DATA[ac] = sat(dsnu[ac],16)
```

where the truncation operator $Y = trunc(X, n)$ eliminates the n less significant bits of the input X and the saturation operator $Y = sat(X, n)$ limits the input X to $2^n - 1$.

The dead column correction performs a neighbor interpolation using the current as well as the two previous samples. Table 5.8 illustrates the principle of this interpolation. Dead columns can be marked for each sample index ac in a separate buffer $DEAD\_LUT$. If a particular index is considered as unreliable, a '1' has to be written to the corresponding buffer location. A '0' will declare the sample index as correct. The algorithm takes the dead column marking of the current as well as the two previous samples $DEAD\_LUT[ac\text{-}2, ac\text{-}1, ac]$ and modifies the linear corrected samples $PREPRO\_DATA$ according to Table 5.8.

5.6.3 FSL Implementation

The preprocessing is very well suited for a pipelined structure. In reality, the un-preprocessed input data are 16-bit samples and the VPU has to process 983 rows within a period of 1 ms. For the experiments the input data width has been reduced to 4-bit and the memory size has

5.6. Hardware Implementation of a Complex Self-Healing Circuit

Table 5.8: Principle of Dead Column Correction

DEAD_LUT			PREPRO_DATA	
DL[ac-2]	DL[ac-1]	DL[ac]	PREPRO_DATA[ac-1]	PREPRO_DATA[ac]
0	0	0	Unchanged	Unchanged
0	0	1	Unchanged	PREPRO_DATA[ac-1]
0	1	0	(PREPRO_DATA[ac-2]+ PREPRO_DATA[ac])/2	Unchanged
0	1	1	PREPRO_DATA[ac-2]	Unchanged
1	0	0	Unchanged	Unchanged
1	0	1	Unchanged	PREPRO_DATA[ac-1]
1	1	0	PREPRO_DATA[ac]	Unchanged
1	1	1	Unchanged	Unchanged

been reduced to 64 rows, so that the circuit was easier to handle. Both parameters are just a matter of scale.

The linear correction in FSL is shown in Figure 5.22. The circuit consists of a three stage pipeline. In the first stage $R1$, the saturation check is performed. Each raw sample $UNPREPRO(ac)$ is compared with the corresponding entry $SAT\_LUT(ac)$. Both the result of this check and the raw input data are stored in register $R1$.

The second stage performs the column non-uniformity correction. The input $UNPREPRO(ac)$ is multiplied by $CRNU\_LUT(ac)$, truncated and limited to 4-bit. The result is stored in register $R2$ together with the saturation information from the previous stage, which simply accompanies the data.

The third stage considers the dark non-uniformity correction by adding $DSNU\_LUT(ac)$ and limiting the result to 4-bit. Depending on the saturation check, either the result of the DSNU correction or the constant SATURATE is stored in register $R3$.

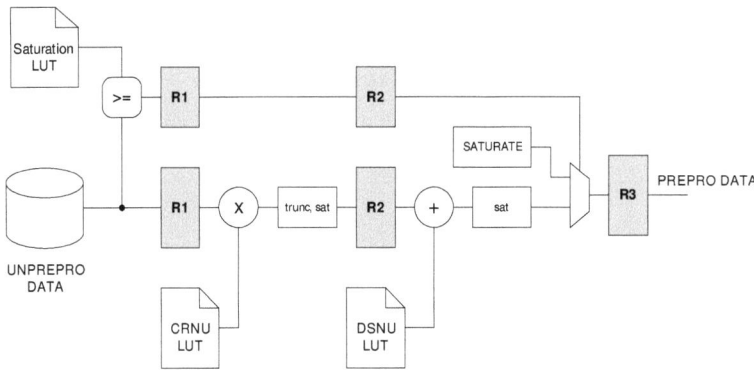

Figure 5.22: Implementation of the Linear Correction

The dead column correction in Figure 5.23 is more complex as the output depends on the value of previous samples. Since the circuit uses feedback elements, phase inverters have to

113

be inserted to ensure that all input data is applied in the correct code phase. The dead column correction consists of two pipeline stages $R4$ and $R5$, which hold $PREPRO\_DATA[ac\text{-}1]$ and $PREPRO\_DATA[ac\text{-}2]$, respectively. The input values to these registers depends on the content of the dead column look-up table, which is used to select the appropriate input via multiplexers.

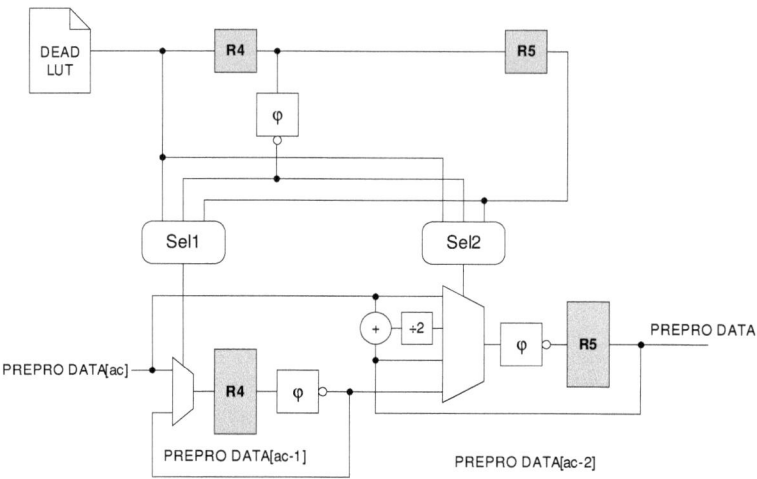

Figure 5.23: Implementation of the Dead Column Correction

The conversion from the synchronous description to the FSL implementation was performed manually and without any assistance by tools, in order to have full control of the structure of the resulting circuit. Also, no radical optimizations for speed or area were performed.

5.6.4 Self-Healing GAIA Algorithm

The GAIA algorithm was transformed to a self-healing circuit following the steps described in section 4.2.8. This was done manually for the same reason as described above for the conversion to an FSL circuit.

Application

Each combinational logic block was replaced by its respective SHC. The 4-bit full adder needed in both the linear correction and the dead column correction was built fine-granular, i.e. the adder itself is composed of basic SHC gates (AND, OR, etc.). This provides more reconfiguration possibilities and a higher fault tolerance but also requires a high amount of reconfiguration inputs, namely 34 per 4-bit adder. To reduce the total amount of reconfiguration inputs all other SHCs (multiplier, multiplexer, etc.) were designed coarse granular, i.e. the nominal and redundant multiplier is designed as one single SHC. This implementation allows to repair faults on any position within the GAIA preprocessing algorithm. Between all pipeline (register) stages an acknowledge switch was implemented.

5.6. Hardware Implementation of a Complex Self-Healing Circuit

Control Environment

The nominal and redundant circuit paths (pipelines) are completely separated, so each pipeline has its own source and sink RAMs. The sink stores all consistent data appearing at the output of the Dead Column Correction (PREPRO_DATA). Together with each token the reconfiguration status is stored, i.e. whether a reconfiguration was required to generate this token.

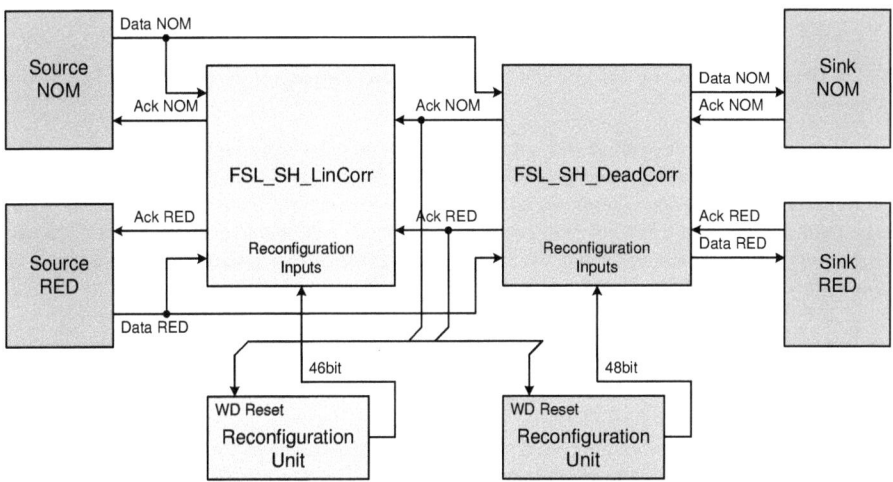

Figure 5.24: Self-Healing GAIA Implementation

Reconfiguration Unit

A dedicated reconfiguration unit was spent for each algorithm, i.e. one for the Linear Correction and a second one for the Dead Column Correction (Figure 5.24). The watchdogs in these two units are reset by activity on the nominal and redundant acknowledge signals between the two algorithm parts.

Within the reconfiguration unit an asynchronous counter was used as watchdog and a synchronous counter, controlled from the watchdog, as reconfiguration controller. As this approach could lead to rather long reconfiguration times (46-bit reconfiguration vector for the Linear Correction and 48-bit for the Dead Column Correction) the relevant reconfiguration bits were mapped to the LSBs of the counter output, so that a reconfiguration could be performed within a feasible time.

5. Analysis, Simulations and Experimental Results

Fault Injection Environment

Saboteurs were implemented at three important locations in the circuit:

- Handshake path: at the nominal and redundant acknowledge signal in the Linear Correction between the first ($R1$) and the second ($R2$) register

- Register input: in the Linear Correction at the nominal and redundant input of register $R2$

- Register output: in the Linear Correction at the output of the nominal and redundant register $R1$

Although the resource effort did not allow to perform exhaustive fault injection experiments with this implementation, it was possible to verify the correct functionality of all basic concepts: A fault in the acknowledge signal can be repaired by a reconfiguration in the *AcknowledgeSwitch*, the fault at the register output (i.e. input to subsequent logic implemented as SHCs) requires a reconfiguration in the SHC and the fault at the register input requires a reconfiguration in both the SHC and the *AcknowledgeSwitch*.

5.6.5 Results

The hardware implementation of the Self-Healing GAIA algorithm worked correctly and basically proved the suitability of the self-healing concept for complex circuits. Fault-injection experiments with single faults showed that the injected permanent fault can be recovered by autonomous re-routing of the application. The reconfiguration controller was a simple counter, i.e. it was a stochastic reconfiguration, and the effects described in section 5.4.2 due to acknowledge toggling were observed. If the reconfiguration was performed manually by applying the correct (manually determined) reconfiguration pattern, the circuit output was always correct.

Resources

Table 5.9 summarizes the resource occupation (number of 4-input LUTs in a Xilinx Virtex-4) as listed in the synthesis reports, as well as the relation to the non-fault tolerant FSL circuit.

Table 5.9: Resource Comparison

	resources	relation
Synchronous GAIA	35	5%
FSL GAIA (reference)	755	100%
SH-GAIA	1565	207%
Reconfiguration Unit (RU)	1865	247%
SH-GAIA incl. RU	3430	454%

The synchronous implementation requires only 5% resources compared to the FSL implementation, which is obvious, as the FPGA is optimized for synchronous architectures. Due to the transformation to a self-healing circuit the logic resources of the FSL implementation are doubled. The acknowledge switches add some additional logic so that the resource occupation

5.6. Hardware Implementation of a Complex Self-Healing Circuit

of the self-healing circuit was expected to be slightly more than twice of the resources of the standard FSL implementation. The synthesis tool reported 207% with respect to the non-fault tolerant GAIA implementation, which sounds reasonable.

The implemented reconfiguration units were asynchronous ripple carry counters, which were more complex than the algorithm of the GAIA application and dominated the total amount of resources. Experiments with counters based on Linear-Feedback Shift Registers (LFSR) resulted in significantly lower values.

Measurements

Figure 5.25 shows the oscilloscope plot of an autonomous reconfiguration of the SH-GAIA algorithm. After the reset is de-asserted (upper waveform, channel 1) the command controller starts to apply source data. The lower waveform (channel 4) shows the acknowledge signal. Each event on the acknowledge signal corresponds to one token shifted through the pipeline. First, three tokens are processed, then a stuck-at-1 fault is injected which holds the acknowledge signal in the high state and causes a deadlock. The watchdog expires several times (not visible in the plot) and after approximately 6ms a working configuration is found. The circuit resumes operation as can be seen on the activity of the acknowledge signal, indicating that the remaining tokens are processed after the reconfiguration.

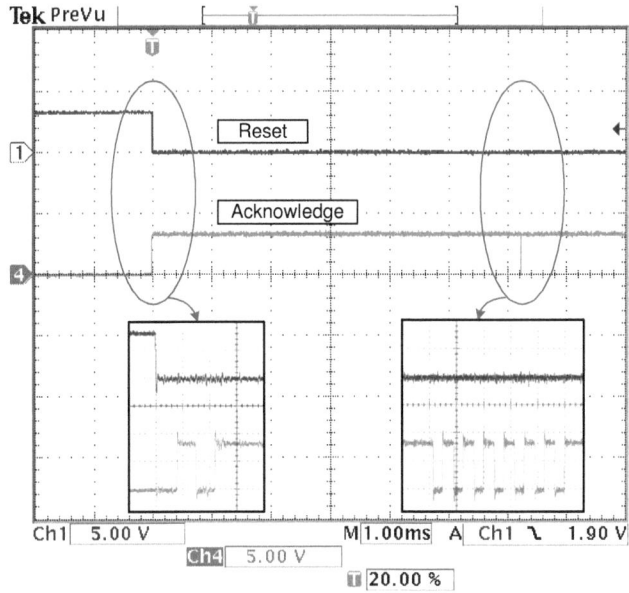

Figure 5.25: Measurement of a Reconfiguration

5.7 Reliability Analysis of Self-Healing Circuit

5.7.1 Introduction

This section presents a reliability estimation for the circuit described in section 5.6, in order to get an impression about the effectiveness of the self-healing concept in terms of reliability. Traditional models for reliability analysis, such as the MIL-HDBK-217 [16], only account for the hardware failure rate of components, and any single fault that causes the component to be out of specification is considered as a total component failure. This means that for programmable integrated components, like FPGAs and ASICs, any applied fault tolerance mechanism that influences the implementation of the circuit in a way that faults could be tolerated, e.g. XTMR [13], is not considered, and thus would not increase the analyzed reliability.

Due to the self-healing concept the circuit is composed from distributed active redundant elements which in total results in a circuit being tolerant against a certain number of faults and fault combinations. In order to compare the reliability of the self-healing implementation with the standard FSL implementation, the methodology of *reliability block diagrams* as usually used on board or system level was applied to our circuits on gate level. For each element used in our circuit (e.g. SHC, register, Acknowledge Switch, etc.) a Matlab model was established which allows determining its reliability considering also possible reconfigurations. For these elements the *signals* (input interfaces of SHCs and registers, internal signals of SHCs, reconfiguration inputs), *switches*, *registers* and *logic* were considered to be either defective or correct, each of these *sub-elements* having assigned a specific failure rate. The model considers all permutations of the involved sub-elements and thus also multiple faults. In the case that a reconfiguration input is defective, the worst case is considered by assuming that the reconfiguration will not be successful. As we know from section 5.3.2, in reality there is a 50% probability that this input is stuck at the intended value, and thus will allow the reconfiguration to take place. For registers an offset (control logic for handshake) and a variable part of the failure rate (depending on the number of inputs) has been considered.

While in the standard for discrete components failure rates were determined based on statistical drop out, no values are available for e.g. an AND gate implemented in an FPGA. We therefore guessed values in a reasonable magnitude and analyzed the reliability for a wide range of the failure rate. In Table 5.10 the used values for the analysis are listed, unless varied over a range. The failure rates (FR) are given in *failures in 10^9 hours [FIT]*.

Table 5.10: Parameters for Reliability Analysis

FR of a signal	0.0005	FIT
FR of a switch	0.1	FIT
FR of a 2-input logic	0.1	FIT
FR of register (offset)	1	FIT
FR of register (variable, per input)	0.01	FIT

The failure rate λ is assumed to be constant, and the resulting reliability can be calculated with the formula $R(t) = e^{-\lambda * t}$. A constant failure rate corresponds to the "normal life" period in the bathtube curve. However, this might not be fully true: (i) for long mission times of tenths of years the wear-out period of the bathtube curve could be reached, where the failure

5.7. Reliability Analysis of Self-Healing Circuit

rate is increasing and (ii) permanent faults which are not caused by hardware defects but e.g. due to radiation might not follow the constant failure rate model. It is assumed that (i) can be avoided by choosing appropriate components. For (ii) the assumption will still hold as long as the occurrence is random, which is usually the case for SEUs.

5.7.2 GAIA Reliability Estimation

To compare the reliability of the GAIA algorithm implementation (see section 5.6) it was first determined which of the parameters listed in Table 5.10 influence the results most. For the detailed analysis later on only the critical parameters were varied. Figure 5.26 shows the gain in reliability for an 8-bit (left plots) and a 16-bit (right plots) self-healing GAIA algorithm compared to a standard FSL implementation. Note that the analysis does not consider the reconfiguration unit, but assumes that the GAIA circuit can be reconfigured. This would be equivalent with a reconfiguration unit having a failure rate of zero. The failure rates for switches and logic were varied over a wide range up to values which seem unrealistic for the hardware failure rate for functions in integrated circuits. However, if also other causes for faults are considered, such as radiation, we might reach even such high values. The lower plots show the magnified area for the section with low failure rates. A mission time of 20 years was assumed. Except of the varied parameters, all values were chosen as listed in Table 5.10. The surface being above the grey plane means that the self-healing implementation achieves a better reliability than the FSL circuit, below it would be worse.

It can be seen that the gain in reliability is significantly influenced by the width of the input vectors (top left vs. top right plot). The reason is, that with a SHC having two reconfiguration inputs, the redundant blocks become larger for higher widths, i.e. each reconfiguration input switches more signals. Thus in case of a fault requiring a reconfiguration, a large amount of resources is lost. If the switches have a high failure rate, they dominate the total reliability, which then might get lower than that of the non-redundant FSL implementation. Assuming that the logic will be more complex and thus have a significantly higher failure rate than the switches, the self-healing implementation provides a significant reliability improvement. With the chosen failure rates the influence of the signals and registers is negligible.

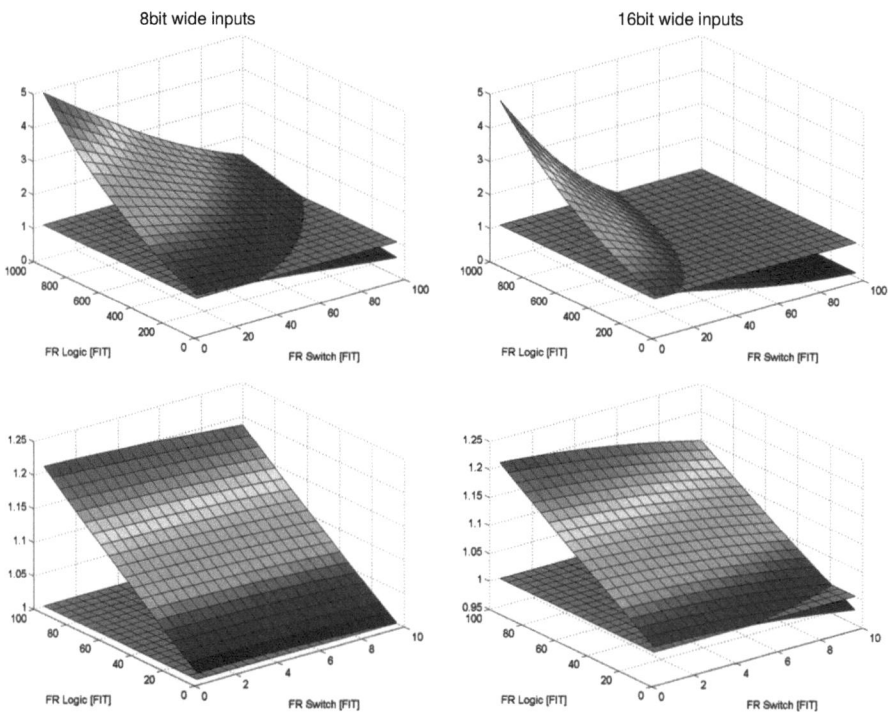

Figure 5.26: Reliability gain of SH vs. FSL GAIA implementation with varying failure rates of switches and logic for 8bit and 16bit wide inputs

5.8 Comparison with State-of-the-Art Methods

In this section the self-healing approach is compared with a duplex-system (1-out-of-2) and a TMR system (1-out-of-3) with respect to multiple fault tolerance.

5.8.1 Duplex System

A duplex system consists of two parallel paths. For simplification it is assumed that, as soon as one path is affected by a fault (**F**alse result), the other one takes over (leading to a **T**rue result). Any detection or switching logic is neglected here.

Already with the first fault the two outputs will disagree, so that the probability for a correct result is 50%. For each subsequent fault the probability of survival decreases by factor $\frac{1}{2}$, which is the probability that the same path is affected again and the other one thus remains fault free.

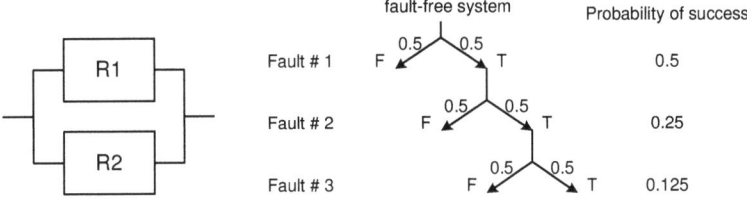

Figure 5.27: Probability of Success for Duplex System

5.8.2 TMR System

A TMR system consists of three parallel paths and the majority is taken as result. The majority voter is not considered in the calculations.

The first fault does not influence the result, as with two working paths a correct majority can be found. A second fault can only be tolerated if the same path is affected, otherwise the output will be wrong. Thus, for each subsequent fault the probability of survival decreases by factor $\frac{1}{3}$.

5.8.3 Comparison with Self-Healing Approach

As can be seen in Table 5.11, the self-healing approach provides significantly higher fault tolerance than a duplex or TMR system. Note that the values for the self-healing approach only consider the pipeline reconfiguration, but not the SHCs. In a typical application the logic functions (implemented in SHCs) will dominate the complexity and amount of resources, and the "administrative logic" such as the pipeline register stages will only have a minor impact. Since SHCs are highly tolerant against multiple faults (refer to section 5.3) the total fault tolerance might thus be even significantly higher than that of the pipeline. As this heavily depends on the logic functions, pipeline length, etc. no such evaluation has been performed within this thesis.

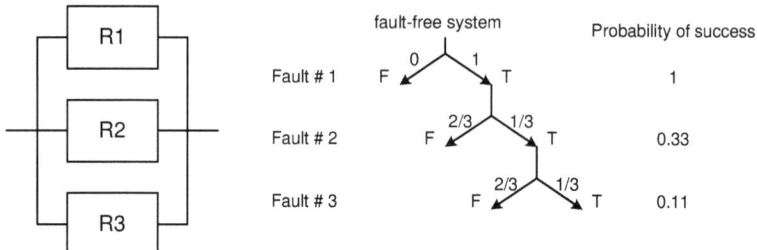

Figure 5.28: Probability of Success for TMR System

Table 5.11: Comparison of Self-Healing Approach with Duplex and TMR Systems

No. of Faults	probability of success			Overhead		
	Duplex	TMR	SH-Approach	Duplex	TMR	SH-Approach
1	50%	100%	100%			
2	25%	33.3%	82.5%	100%	200%	115%..240%
3	12.5%	11.1%	56.7%			

The resource overhead given in Table 5.11 is calculated as $(R_{ft} - R_0)/R_0$, where R_0 are the resources of the simplex system and R_{ft} the resources of the fault tolerant system. If the logic for detection, voting and switching is neglected, the total resources of a duplex system increase by a factor of 2, for a TMR system by a factor of 3 compared to a simplex (non-redundant) system. Assuming 10% of the resources being related to registers and 90% being related to logic, the resource effort for the self-healing approach increases by a factor of approximately 3.3 (see section 4.6).

The resource effort for the reconfiguration units described in section 5.4.2 will not be significant, as they are basically lookup-tables with latches at the outputs that store the configuration bit of each reconfiguration input. The overhead is assumed to be in the same order as the voting and switching logic in a TMR system.

Overall, the relative resource overhead of the self-healing concept for a typical application will be between that of a duplex and a TMR system for a coarse granular implementation, and slightly above that of a TMR system for a fine granular implementation.

Basically the self-healing approach could be applied to duplex and TMR systems to make them repairable and thus increase the fault tolerance even more. However, for asynchronous circuits this is not as easy as for synchronous logic and requires particular techniques [102]. Of course, the resource overhead will increase significantly with this approach.

The overhead of the method using reconfigurable logic blocks [69] is similar for basic gates (230%), but considerably lower for more complex circuits (e.g. < 40% for an 8-bit ALU). However, these values do not include the effort for diagnosis and the approach does not cover the connections between the implemented functions.

While the approach based on a self-healing asynchronous linear array [101] reports lower overheads for fault tolerance up to two faults (e.g. 74% for a 64-bit adder being tolerant against one fault), the overhead increases significantly and exceeds the values of the self-healing

5.9. Summary and Discussion

approach described in this thesis for multiple fault tolerance (e.g. 307% for an 8-bit adder being tolerant against three faults, 387% for four faults). Although a k-fault tolerant array can guarantee the recovery from k faults, this method might soon become unattractive for a higher number of faults due to the enormous resource effort. Our self-healing approach does not guarantee a k-fault tolerance, but provides a high probability for it at much lower resource overhead.

5.9 Summary and Discussion

This chapter presented the results of various simulations and hardware fault injection experiments performed on different levels of the self-healing concept. In general, it was shown that the concept is able and suitable to improve the fault tolerance of FSL circuits. In particular, the concept provides also a significant probability for multiple fault tolerance and is advantageous compared to duplex and TMR systems in this respect.

The presented self-healing approach is beneficial in several respects compared to the methods described in chapter 3:

- It is not restricted to FPGAs but also applicable to ASICs.

- The concept is applied on gate level, and the FSL circuits can use standard component libraries for integrated circuits.

- The actual circuit configuration can easily be determined via the reconfiguration vector, so the circuit's behavior is predictable even if stochastic reconfiguration-algorithms are used.

- SHCs also cover faults in the interconnections and can implement any kind of circuits.

- The different configurations do not need to be generated during design time.

- Since the concept is based on asynchronous circuits, a different timing due to a modified configuration does not influence the circuit functionality.

- The concept provides inherent state-recovery. In general, FSL circuits deadlock when permanent faults occur but keep the data. The circuit continues autonomously after the reconfiguration without loss or corruption of data.

- Most methods are restricted to FPGAs and use the external configuration interface, which is a single-point of failure. The SHCs are distributed across the whole circuit and are controlled by only a few signals without particular protocol behind. Thus no central configuration interface is needed.

Some important general observations obtained during the simulations and experiments are summarized below.

5. Analysis, Simulations and Experimental Results

5.9.1 Fault Frequency

Due to the handshake mechanism, effects occurring in the pipeline might not be visible immediately. Depending on the location within the pipeline, a permanent fault will not instantaneously lead to a deadlock, but the data and the handshakes could still work for some cycles, until the latent fault blocks the handshake protocol.

On the one hand, this delays the fault detection by some time which depends on the pipeline length, because the fault effect must propagate through the pipeline until all handshakes are stopped. On the other hand, this might have an impact on the functionality of the reconfiguration unit. As shown in chapter 4, each fault ends up in a characteristic pattern of the handshake signals and phase detector status. However, this pattern is only visible if the fault has completely settled in the pipeline. If another fault occurs too early, i.e. while the effects of the previous fault are still propagating in the pipeline, the resulting pattern might be interpreted incorrectly and thus a wrong reconfiguration rule could be applied. Consequently, this results in a maximum fault frequency, which depends on the pipeline length and the reconfiguration speed. The reconfiguration of one fault must be finished before the next fault occurs in the pipeline.

We assume repair in the millisecond range, while the rate of permanent faults is orders of magnitude higher (e.g. in [11] a failure rate of 7.73 failures per 10^9 hours is reported for the RTAX-S antifuse FPGA, corresponding to a mean-time-to-fail (MTTF) of 1.29E+08 hours (¿14767 years)). Although transient faults can cause deadlocks in FSL circuits [28], even with SEU rates of e.g. 4.0E-03 SEUs per device day in GEO orbit (MTBF of ~ 250 days) as given in [57] for the space qualified floating point DSP SMV320C6701, the MTTF will remain significantly higher than the repair time.

5.9.2 Timing Assumptions

In section 4.2.5 the timing constraints due to the deadlock detector (watchdog circuit) were explained. As stated there, although the watchdog formally violates the unbounded delay model, it is no practical restriction if the timeout is chosen several orders of magnitude higher than the maximum circuit processing time.

For the hardware implementation of the asynchronous circuits an asynchronous component library was established, which can be used for synthesis in standard FPGAs (for the experiments a Xilinx Virtex-4 FPGA was used). These components assume delay insensitivity on gate level. Although this is not guaranteed by the design, the experiments showed that the assumption seems to be fulfilled in most of the cases.

Splitting the data and acknowledge path requires additional timing restrictions (see section 4.3.5), which cannot be solved by simply using delay insensitive components. In particular the relative timing between the nominal and redundant pipeline path needs to be considered in the circuit routing. As no such timing definition was performed for the circuits used for the hardware experiments, it is likely that timing problems occurred. This would explain some effects that were seen during the experiments (e.g. as described in section 5.5).

5.9.3 Mis-alignment of Nominal and Redundant Path

The considerations in this thesis assume permanent faults, i.e. faults that appear at any time and remain forever. A particular phenomenon has been observed if a fault lasts for a long time

5.9. Summary and Discussion

(long enough so that a reconfiguration takes place and brings the circuit back to operation), but not forever, i.e. it disappears later on.

Depending on the fault location, circuit granularity and applied reconfiguration, it might be the case that only one of the two pipeline outputs provides new tokens, and the other output is stopped. If the fault disappears, also this output could start to produce tokens again. For a counter, e.g., this could lead to two working outputs with different values. Such a behavior must be handled on application level.

5.10 Annex

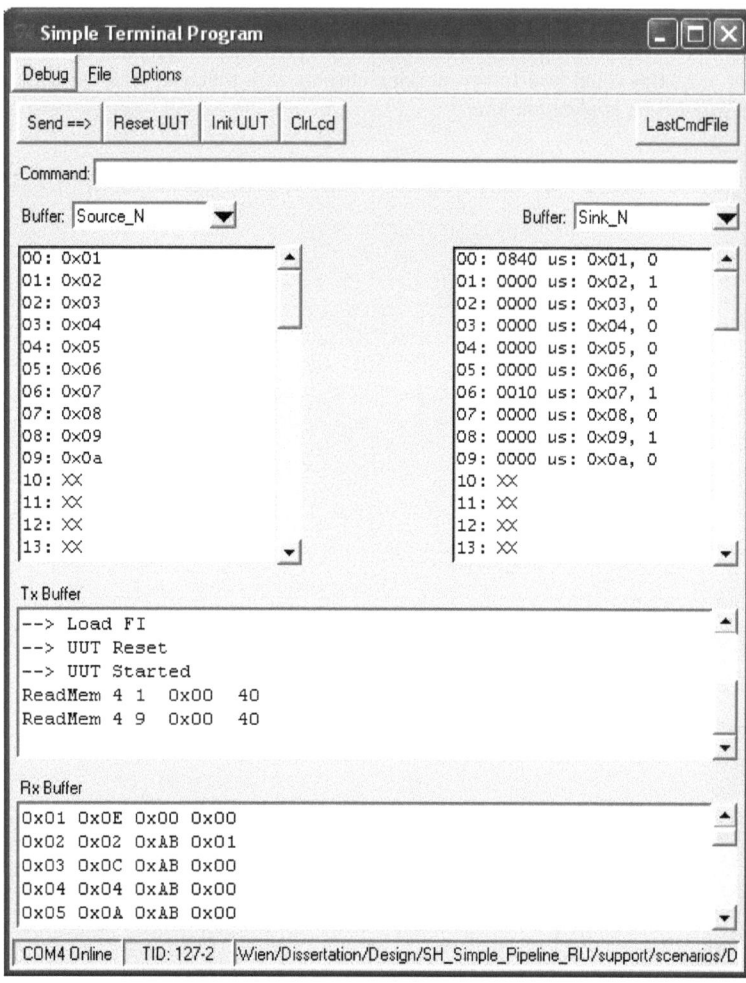

Figure 5.29: GUI for Controlling the Hardware Fault Injection Experiments

5.10. Annex

Listing 5.6: Example of a Stimulus File for HW Fault Injection

```
1  InitLog .\\result\\HW_Sim_sequence_2-to-2-faults_1-1.txt
2  LogDetail OFF
3  LogSummary ON
4  --Flush sink buffers
5  Reset
6  WrMem 0, 0x0 0x0 0x0 0x0 0x0 0x0 0x0 0x0 0x0 0x0 0x0 0x0 0x0 0x0
7           0x0 0x0 0x0 0x0 0x0 0x0
8  WrMem 8, 0x0 0x0 0x0 0x0 0x0 0x0 0x0 0x0 0x0 0x0 0x0 0x0 0x0 0x0
9           0x0 0x0 0x0 0x0 0x0 0x0
10 SetLength 20
11 ResetFI
12 Start
13 SrcDelay 20
14
15 -- 1-1) NEW SEQUENCE --- 2 faults applied, sequence: 30-29
16 Reset
17 WrMem 0, 0x1 0x2 0x3 0x4 0x5 0x6 0x7 0x8 0x9 0xA 0xB 0xC 0xD 0xE
18          0xF 0x1 0x2 0x3 0x4 0x5
19 WrMem 8, 0x1 0x2 0x3 0x4 0x5 0x6 0x7 0x8 0x9 0xA 0xB 0xC 0xD 0xE
20          0xF 0x1 0x2 0x3 0x4 0x5
21 SetLength 20
22 ResetFI
23 -- enable WD
24 FaultInjection R C 1 0 ON 0 0
25 -- Injecting 1. fault
26 -- stuck-at-1 on 137 (Reg5.PassRed) in cycle 2 (absolute), offset
27    2 clock cycles
28 FaultInjection A F 2 137 ON 1 2
29 -- Injecting 2. fault
30 -- stuck-at-1 on 103 (Reg4.PassRed) in cycle 9 (absolute), offset
31    2 clock cycles
32 FaultInjection A F 9 103 ON 1 2
33 LoadFI
34 TestID 1-1
35 Reset
36 Start
37 RdMem4 1, 0x00, 80
38 RdMem4 9, 0x00, 80
39 LogData
```

Listing 5.7: Rules of 2-Input SHC Model

```
1  if SHC1.ValOutNom ~= 'X'
2    if (SHC1.ReconfNom == 0)
3      if SHC1.ValInNom ~= 'X'
4        SHC1.ValOutNom = SHC1.ValInNom;
5        SHC1.PhiOutNom = SHC1.PhiInNom;
6      end
7    else
8      if SHC1.ValInRed ~= 'X'
9        SHC1.ValOutNom = SHC1.ValInRed;
10       SHC1.PhiOutNom = SHC1.PhiInRed;
11     end
12   end
13 end
```

5.10. Annex

Listing 5.8: Rules of Register Model

```
if Reg1.ValInNom ~= 'X'
  % do not propagate a defect-marker
  if SHC1.ValOutNom ~= 'X'
    Reg1.ValInNom = SHC1.ValOutNom;
    Reg1.PhiInNom = SHC1.PhiOutNom;
  end
end
if Reg1.ValInRed ~= 'X'
  % do not propagate a defect-marker
  if SHC1.ValOutRed ~= 'X'
    Reg1.ValInRed = SHC1.ValOutRed;
    Reg1.PhiInRed = SHC1.PhiOutRed;
  end
end

% transparent: phi_in != phi_out, phi_out=pass

if (Reg1.PhiInNom ~= Reg1.PhiOutNom)
  if (Reg1.PhiOutNom == Reg1.PassNom) & (Reg1.PassNom ~= 9)
    % transparent
    Reg1.ValOutNom = Reg1.ValInNom;
    Reg1.PhiOutNom = Reg1.PhiInNom;
    % set done signal accordingly
    Reg1.DoneNom = Reg1.PhiOutNom;
  end
end

if (Reg1.PhiInRed ~= Reg1.PhiOutRed)
  if (Reg1.PhiOutRed == Reg1.PassRed) & (Reg1.PassRed ~= 9)
    % transparent
    Reg1.ValOutRed = Reg1.ValInRed;
    Reg1.PhiOutRed = Reg1.PhiInRed;
    % set done signal accordingly
    Reg1.DoneRed = Reg1.PhiOutRed;
  end
end
```

Listing 5.9: Rules of Acknowledge Switch Model

```
% AS between Reg1 and Reg2
AS12.AckInNom = Reg2.DoneNom;
AS12.AckInRed = Reg2.DoneRed;

if (AS12.ReconfNom == 0)
  AS12.AckOutNom = AS12.AckInNom;
else
  AS12.AckOutNom = AS12.AckInRed;
end
% AS between Reg1 and Reg2
if (AS12.ReconfRed == 0)
  AS12.AckOutRed = AS12.AckInRed;
else
  AS12.AckOutRed = AS12.AckInNom;
end
```

Chapter 6
Conclusion and Outlook

In this thesis a self-healing concept based on asynchronous circuits is presented. It uses a redundant pipeline as basic circuit structure and replaces the combinational logic by Self-Healing Cells (SHC), which provide the possibility of reconfiguration. The concept utilizes the inherent properties of the asynchronous design style FSL, in particular the fail-stop behavior, which helps to reduce the resource effort for fault diagnosis. The reconfigurable elements are distributed over the whole circuit, so the concept is able to tolerate also multiple faults, which makes it particularly interesting for long mission times. The granularity, and thus the fault tolerance, is variable and can be defined by the designer as needed to achieve the reliability goal for the application.

The objectives as defined in section 1.2 have essentially been accomplished:

- *Recover from multiple permanent faults and errors*: achieved; the fault tolerance determined from the experiments is $> 83\%$ for double faults and nearly 60% for triple faults; further improvement is possible with more intelligent reconfiguration units

- *occurring in integrated circuits*: achieved; experiments have been performed with circuits implemented in a standard FPGA, but the concept can be applied also to ASICs; the concept covers interconnects, combinational and sequential logic

- *within a predictable timing*: achieved; the duration of a reconfiguration using the investigated algorithms is limited to an upper boundary which depends on the pipeline length

- *with deterministic measures*: achieved; the symptoms occurring in case of a permanent fault lead to a clear reconfiguration pattern; by knowing this pattern, the actual circuit structure is unambiguously defined

- *by autonomous reconfiguration*: achieved; a watchdog circuit monitors the circuit's activity and triggers the reconfiguration process during runtime

- *transparent to the application*: achieved; due to the inherent fail-stop behavior of FSL logic, the application is autonomously stopped for the reconfiguration process; after the reconfiguration the circuit starts working again autonomously without external interaction by the application or the user

- *using existing processes and standard libraries*: achieved; the concept is applied on gate level, and although optimizations with a dedicated library are assumed to be possible, it was shown that the implementation works even with standard elements

The main focus of the thesis was to handle faults that occur during operation and lifetime, i.e. the application was working correctly before the fault occurred. In fact, the concept can also be used to increase the yield, as after switch-on it will immediately change the circuit's routing so that it starts working. However, no detailed investigation has been performed with respect to this application.

The apparent drawback of asynchronous circuits is the initial overhead compared to synchronous implementations. The presented approach is thus not proposed as replacement for fault tolerance mechanism in the synchronous domain. However, if asynchronous circuits are being used, the concept is an appropriate method to increase the fault tolerance significantly by introducing a relative overhead comparable to a TMR system. The author is aware that the simulations and experiments performed do not provide adequate coverage to conclude on the fault tolerance for any arbitrary system, but the results justify the suitability of the presented approach for tolerating multiple faults.

Although the concept is beneficial in several points compared to other methods of circuit reconfiguration, some issues have been identified, which need further investigation. The following list shall therefore be seen as some kind of a roadmap for the continuation of research in this area.

- Basically, the reconfiguration process performs without loss or corruption of data and internal states are recovered. However, FSL logic is not hazard-free any more in presence of permanent faults, which could lead to wrong data being propagated in the circuit. It is assumed, that such effects must be handled on application level.

- In this thesis two independent data sources and sinks were assumed. The circuit was defined to be working, if at least one of the two outputs showed activity. The approach itself, however, does not include methods to identify which of the two outputs is the correct one. Thus, the management of the redundant pipeline must be solved.

- The simulations and experiments revealed that it is not sufficient to just establish any valid data and acknowledge path, but some timing constraints need to be fulfilled. A comprehensive and detailed study of the path timings could not be performed due to insufficient tool support, while a manual inspection would cause prohibitive efforts. Although the few available timing results confirmed the theoretical predictions very well, a more detailed investigation of both the working and the failing configurations would have yielded an even higher confidence in the predictions, or might also have revealed interesting new insights. With the availability of newer tool versions and/or circuits optimized for this scope such investigations will become possible.

- Currently a fixed watchdog timeout is used. This could be improved, e.g. to start the watchdog with each new token provided by the source, to avoid timeouts due to pauses of data inputs.

- Random algorithms cause problems during operation but would increase the number of possible configurations. It should be investigated, if e.g. a "test mode" could be added to establish a new configuration without disturbing the actual tokens.
- As transient faults are more likely to occur than permanent faults, the application of additional measures to increase also the robustness against transient fault should be investigated.
- In scope of this thesis the concept was only applied using the FSL coding. As long as the "fail-stop" behavior is guaranteed, basically also other codes (e.g. NCL) could be used. However, the manifestation of fault effects might be different and has to be carefully assessed.
- In principle the concept can also be used for synchronous circuits. However, while the fail-stop behavior is "for free" in FSL, particular means are needed in synchronous circuits which add additional complexity and resources. Overall, it might still be an interesting option, and research in this area is worth to be done.

Bibliography

[1] Melvin A.Breuer, Sandeep K.Gupta, and T.M.Mak. Defect and error tolerance in the presence of massive numbers of defects. *Journal of IEEE Design and Test of Computers, Vol.21, Iss.3*, 2004.

[2] Algirdas Avizienis, Jean-Claude Laprie, Brian Randell, and Carl Landwehr. Basic concepts and taxonomy of dependable and secure computing. In *IEEE Transactions on Dependable and Secure Computing*, 2004.

[3] Babak Rahbaran. *An Experimental Comparison of Robustness between Synchronous and Asynchronous Logic Design*. PhD thesis, Vienna University of Technology / Department of Computer Engineering, 2005.

[4] Shekhar Borkar. Designing reliable systems from unreliable components: The challenges of transistor variability and degradation. *IEEE Micro*, 25:10–16, November 2005.

[5] David Brodrick, Anwar Dawood, Neil Bergmann, and Melanie Wark. Error detection for adaptive computing architectures in spacecraft applications. In *Proceedings of the 6th Australasian Computer Systems Architecture Conference (ACSAC 2001)*, 2001.

[6] Stephen Brown, Jonathan Rose, and Zvonko G. Vranesic. A detailed router for field-programmable gate arrays. *IEEE Transactions on Computer-Aided Design, Vol.11*, 1992.

[7] Actel Corporation. Actel axcelerator antifuse devices. http://www.actel.com/products/axcelerator/default.aspx (02 Dec 2011).

[8] Actel Corporation. RTSX-S radtolerant FPGAs. Datasheet 517151-11/11.04 (v2.2), 2004.

[9] Actel Corporation. FPGA reliability and the sunspot cycle. Technical Report 55900103-0.9.09, 2009.

[10] Altera Corporation. Altera stratix ii device handbook, volume 1. Handbook SII5V1-4.5, April 2011.

[11] Microsemi Corporation. Reliability report. Report 51000001-9/8.11, August 2011.

[12] Xilinx Corporation. Triple module redundancy design techniques for virtex FPGAs. Application Note XAPP197, November 2001.

[13] Xilinx Corporation. Xilinx TMRTool. Product Brief PN2404, 2009.

[14] Martin Delvai, Gottfried Fuchs, Thomas Handl, Wolfgang Huber, and Andreas Steininger. Design of an asynchronous microprocessor with four-state logic. In *Proceedings of Austrochip 2005*, 2005.

[15] Martin Delvai and Andreas Steininger. Asynchronous logic design - from concepts to implementation. *The 3rd International Conference on Cybernetics and Information Technologies, Systems and Applications - Volume 1*, January 2006.

[16] Department of Defense. Military Handbook - Reliability Prediction of Electronic Equipment. MIL-HDBK-217F, Notice 2, 1995.

[17] P.E. Dodd, M.R. Shaneyfelt, J.A Felix, and J.R. Schwank. Production and propagation of single-event transients in high-speed digital logic ics. In *IEEE Transactions on Nuclear Science, Vol. 51, No. 6*, pages 3278–3284, December 2004.

[18] Abderrahim Doumar and Hideo Ito. Detecting, diagnosing and tolerating faults in sram-based field programmable gate arrays: A survey. *IEEE Transactions on Very Large Scale Integration (VLSI) Systems, Vol.11, No.3*, 2003.

[19] Abderrahim Doumar, Satoshi Kaneko, and Hideo Ito. Defect and fault tolerance FPGAs by shifting the configuration data. In *Proceedings of the IEEE International Symposium on Defect and Fault Tolerance in VLSI Systems (DFT'99)*, November 1999.

[20] EADS Astrium. VPU Algorithms Requirements Specifications, GAIA.ASF.SP.PLM.00073. Invititation to Tender for GAIA VPU, 2006.

[21] John Emmert, Charles Stroud, Brandon Skaggs, and Miron Abramovici. Dynamic fault tolerance in FPGAs via partial reconfiguration. In *Proceedings of the 2000 IEEE Symposium on Field-Programmable Custom Computing Machines (FCCM'00)*, Washington, DC, USA, 2000.

[22] European Space Agency. Space Science - BepiColombo Overview. http://www.esa.int/science/bepicolombo (05 Dec 2011).

[23] European Space Agency. The GAIA Mission. http://sci.esa.int/gaia/ (05 Dec 2011).

[24] European Space Components Information Exchange System (ESCIES). ESA Radiation. https://escies.org/ReadArticle?docId=227 (06 Dec 2009).

[25] A.H. Fischer, A. von Glasow, S. Penka, and F. Ungar. Electromigration failure mechanism studies on copper interconnects. In *Proceedings of the IEEE 2002 International Interconnect Technology Conference*, pages 139–141, 2002.

[26] B. Folco, V. Bregier, L. Fesquet, and M. Renaudin. Technology mapping for area optimised quasi delay insensitive circuits. In *Proceedings of the International Conference on Very Large Scale Integration (VLSI-SOC)*, pages 146–151, 2005.

[27] European Cooperation for Space Standardization. Derating - eee components. Standard ECSS-Q-30-11A, 2006.

Bibliography

[28] Werner Friesenbichler. *Effects and Mitigation of Transient Faults in Quasi Delay-Insensitive Logic*. PhD thesis, Vienna University of Technology / Department of Computer Engineering, 2011.

[29] M. Gericota, L. Lemos, G. Alves, and J. Ferreira. A framework for self-healing radiation-tolerant implementations on reconfigurable FPGAs. In *Proceedings of the 10th IEEE Workshop on Design and Diagnostics of Electronic Circuits and Systems (DDECS'2007)*, April 2007.

[30] M. Gericota, L. Lemos, G. Alves, and J. Ferreira. On-line self-healing of circuits implemented on reconfigurable FPGAs. In *13th IEEE International On-Line Testing Symposium (IOLTS 2007)*, July 2007.

[31] Manuel G.Gericota, Gustavo R.Alves, Miguel L.Silva, and Jose M.Ferreira. Active replication: Towards a truly sram-based FPGA on-line concurrent testing. In *Proceedings of the Eigth IEEE International On-Line Testing Workshop (IOLTW'02)*, 2002.

[32] Manuel G.Gericota, Gustavo R.Alves, Miguel L.Silva, and Jose M.Ferreira. Dynamic replication: The core of a truy non-intrusive sram-based FPGA structural concurrent test methodology. In *Proceedings of the Third IEEE Latin American Test Workshop*, February 2002.

[33] Manuel G.Gericota, Gustavo R.Alves, Miguel L.Silva, and Jose M.Ferreira. Run-time management of logic resources on reconfigurable systems. In *Proceedings of the Design, Automation and Test in Europe Conference and Exhibition (DATE'03)*, 2003.

[34] Manuel G.Gericota, Gustavo R.Alves, and Jose M.Ferreira. Dynamically rotate and free for test: The path for FPGA concurrent test. In *Proceedings of the second IEEE Latin American Test Workshop*, February 2002.

[35] Manuel G.Gericota, Gustavo R.Alves, and Jose M.Ferreira. Robust configurable system design with built-in self-healing. In *Proceedings of the 20th Conference on Design of Circuits and Integrated Systems (DCIS'05)*, November 2005.

[36] Manuel G.Gericota, Gustavo R.Alves, and Jose M.Ferreira. A self-healing real-time system based on run-time self-reconfiguration. In *Proceedings of the 10th IEEE International Conference on Emerging Technologies and Factory Automation (ETFA'2005)*, September 2005.

[37] Christian Gleichner, Tobias Koal, and Heinrich T. Vierhaus. Effiziente Verfahren der Selbstreparatur von Logik. In *Proceedings of the 22. ITG/GI/GMM Workshop Testmethoden und Zuverl"assigkeit von Schaltungen und Systemen*, 2010.

[38] G.M.Swift, D.J.Padgett, and A.H.Johnston. A new class of single event hard errors. *IEEE Transactions on Nuclear Science*, September 1994.

[39] Michael Goessel, Vitaly Ocheretny, Egor Sogomonyan, and Daniel Marienfeld. *New Methods of Concurrent Checking*. Springer, 2008.

Bibliography

[40] Maya Gokhale, Paul Graham, Eric Johnson, Nathan Rollins, and Michael Wirthlin. Dynamic reconfiguration for management of radiation-induced faults in FPGAs. In *Proceedings of the 18th International Parallel and Distributed Processing Symposium (IPDPS'04)*, 2004.

[41] Liu Fang Dai Kui Wang Zhiying Gong Rui, Chen Wei. Modified triple modular redundancy structure based on asynchronous circuit technique. In *Proceedings of the 21st IEEE International Symposium on Defect and Fault-Tolerance in VLSI Systems (DFT'06)*, 2006.

[42] Paul Graham, Michael Caffrey, Michael Wirthlin, Eric Johnson, and Nathan Rollins. Reconfigurable computing in space: From current technology to reconfigurable systems-on-a-chip. In *Proceedings of the IEEE 2003 Aerospace Conference*, 2003.

[43] Steven A. Guccione, Delon Levi, and Prasanna Sundararajan. Jbits: A java-based interface to FPGA hardware. In *Proceedings of the 2nd Annual Military and Aerospace Applications of Programmable Devices and Technologies Conference (MAPLD)*, 1999.

[44] Fran Hanchek and Shantanu Dutt. Node-covering based defect and fault tolerance methods for increased yield in FPGAs. In *Proceedings of the 9th International Conference on VLSI Design*, 1996.

[45] Fran Hanchek and Shantanu Dutt. Methodologies for tolerating cell and interconnect faults in FPGAs. *IEEE Transactions on Computers*, January 1998.

[46] Scott Hareland, Jose Maiz, Mophsen Alavi, Kaizad Mistry, Steve Walsta, and Changhong Dai. Impact of cmos process scaling and soi on the soft error rates of logic processes. *Digest of Technical Papers of the 2001 Symposium on VLSI Technology*, 2001.

[47] S. Hauck. Asynchronous design methodologies: An overview. In *Proceedings of the IEEE, Vol. 83, No. 1*, pages 69–93, January 1995.

[48] Scott Hauck, Gaetano Borriello, Steven Burns, and Carl Ebeling. MONTAGE: An FPGA for Synchronous and Asynchronous Circuits. In *Proceedings of the Second International Workshop on Field-Programmable Logic and Applications (FPL)*, 1992.

[49] Gaetano Borriello Henrik Hulgaard, Steven M. Burns. Testing asynchronous circuits: A survey. *University of Washington, Technical Report 94-03-06*, 1994.

[50] Wei-Je Huang and Edward J.McCluskey. Column-based precompiled configuration techniques for FPGA fault tolerance. In *Proceedings of the 9th Annual IEEE Symposium on Field-Programmable Custom Computing Machines (FCCM'01)*, 2001.

[51] Wei-Je Huang and Edward J.McCluskey. A memory coherence technique for online transient error recovery of FPGA configurations. In *Proceedings of the 2001 ACM/SIGDA ninth international symposium on Field programmable gate arrays*, February 2001.

[52] Wei-Je Robert Huang. *Dependable Computing Techniques for Reconfigurable Hardware*. PhD thesis, Stanford University, Center for Reliable Computing, 2001.

Bibliography

[53] D. Huffman. The synthesis of the sequential switching circuits. In *Journal of the Franklin Institute, Vol. 257, No. 4*, pages 161–190, March 1954.

[54] G. Komoriya H. Pham F. P. Higgins J. L. Lewandowski I. Kim, Y. Zorian. Built in self repair for embedded high density sram. In *Proceedings of IEEE International Test Conference 1998*, 1998.

[55] Cray Inc. Cray xd1 supercomputer. Cray XD1 Datasheet, 2004.

[56] Xilinx Inc. Virtex-4 family overview. Data Sheet DS112 (v3.1), August 2010.

[57] Texas Instruments Incorporated. Qml class v product portfolio. Datasheet SGZM004, 2003.

[58] International Technology Roadmap for Semiconductors. 2010 Update. http://www.itrs.net/Links/2010ITRS/Home2010.htm (08 Jan 2011).

[59] S. Ishihara, Y.Komatsu, M.Hariyama, and M.Kameyama. An asynchronous FPGA using ledr/4-phase-dual-rail protocol converters. In *Proceedings of the International Conference on Reconfigurable Systems and Algorithms (ERSA)*, pages 145–150, 2009.

[60] Wonjin Jang and Alain J. Martin. Seu-tolerant qdi circuits. In *Proceedings of the 11th IEEE International Symposium on Asynchronous Circuits and Systems (ASYNC'05), pp. 156-165*, 2005.

[61] Wonjin Jang and Alain J. Martin. Soft-error robustness in qdi circuits. *Workshop on System Effects of Logical Soft Errors - SELSE 1*, April 2005.

[62] JEDEC. Failure mechanisms and models for semiconductor devices. JEP122F, 2010.

[63] J.F. Ziegler and W.A. Lanford. Effect of Cosmic Rays on Computer Memories. Science, November 1979.

[64] B. Johnson. *Design and Analysis of Fault Tolerant Digital Systems*. Addison Wesley, 1989.

[65] J. Becker K. Paulsson, M. Huebner. Strategies to on-line failure recovery in self-adaptive systems based on dynamic and partial reconfiguration. In *Proceedings of the First NASA/ESA Conference on Adaptive Hardware and Systems (AHS'06)*, 2006.

[66] Cindy Kao. Benefits of partial reconfiguration. Xcell Journal, Fourth Quarter, October 2005.

[67] Fernanda Lima Kastensmidt, Gustavo Neuberger, Luigi Carro, and Ricardo Reis. Designing and testing fault-tolerant techniques for sram-based FPGAs. In *Proceedings of the 1st conference on Computing frontiers (CF'04)*, pages 419–432, New York, NY, USA, 2004. ACM.

[68] T. Koal, D. Scheit, and H. T. Vierhaus. A concept for logic self repair. In *Proceedings of the 12th Euromicro Conference on Digital System Design (DSD)*, pages 621–624, 2009.

[69] Tobias Koal, Daniel Scheit, and Heinrich T. Vierhaus. Selbstreparatur durch regularisierung von logik-schaltungen. In *Proceedings of the 3. GMM/GI/ITG-Fachtagung Zuverl"assigkeit und Entwurf*, 2009.

[70] Tobias Koal, Daniel Scheit, and Heinrich T. Vierhaus. Schwachstellen und engp"asse bei verfahren der selbstreparatur f"ur hochintegrierte schaltungen und systeme. In *Proceedings of the 4. GMM/GI/ITG-Fachtagung Zuverl"assigkeit und Entwurf*, 2010.

[71] Philip Koopman. Elements of the self-healing system problem space. In *Proceedings of the ICSE 2003 Workshop on Software Architectures for Dependable Systems (WADS2003)*, 2003.

[72] H. Kopetz. System failure is the norm, not the exception. In *Keynote, DATE 2008*, 2008.

[73] Hermann Kopetz. *Real-Time Systems: Design Principles or Distributed Embedded Applications*. Kluwer Academic Publishers, 1997.

[74] John Lach, William H.Mangione-Smith, and Miodrag Potkonjak. Low overhead fault-tolerant FPGA systems. *IEEE Transactions on Large Scale Integration (VLSI) Systems*, June 1998.

[75] John Lach, William H.Mangione-Smith, and Miodrag Potkonjak. Enhanced FPGA reliability trough efficient run-time fault reconfiguration. *IEEE Transactions on Reliability, Vol.49, No.3*, 2000.

[76] Christopher LaFrieda and Rajit Manohar. Fault detection and isolation techniques for quasi delay-insensitive circuits. In *Proceedings of the International Conference on Dependable Systems and Networks*, July 2004.

[77] Fernanda L.Kastensmidt, L.Sterpone, Luigi Carro, and M.Sonza Reorda. On the optimal design of triple modular redundancy logic for sram-based FPGAs. In *Proceedings of the Design, Automation and Test in Europe Conference and Exhibition (DATE'05)*, 2005.

[78] Rajit Manohar. A case for asynchronous computer architecture. In *Proceedings of the ISCA Workshop on Complexity-Effective Design*, 2000.

[79] Alain J. Martin. The limitations to delay-insensitivity in asynchronous circuits. In *Proceedings of the sixth MIT conference on Advanced research in VLSI*, pages 263–278, Cambridge, MA, USA, 1990. MIT Press.

[80] Alain J. Martin and Pieter J. Hazewindus. Testing delay-insensitive circuits. *Technical Report Caltech-CS-TR-90-17*, 1990.

[81] John McCardle and Dr. David Chester. Measuring an asynchronous processor's power and noise. *Synopsis Users Group Boston*, 2001.

[82] Dennis McCarty. Incremental place and route speeds debug iterations. *The Syndicated*, 2005.

Bibliography

[83] Scott McMillan and Steve Guccione. Partial run-time reconfiguration using jrtr. In *Proceedings of the The Roadmap to Reconfigurable Computing, 10th International Workshop on Field-Programmable Logic and Applications*, FPL '00, pages 352–360, London, UK, 2000. Springer-Verlag.

[84] John M.Emmert and Jason A.Cheatham. On-line incremental routing for interconnect fault tolerance in FPGAs minus the router. In *Proceedings of the 2001 IEEE international Symposium on Defect and Fault Tolerance in VLSI Systems (DFT'01)*, 2001.

[85] John M.Emmert and Dinesh Bhatia. Incremental routing in FPGAs. In *Proceedings of the 11th Annual IEEE International ASIC Conference 1998*, 1998.

[86] Amitava Mitra, William F. McLaughlin, and Steven M. Nowick. Efficient asynchronous protocol converters for two-phase delay-insensitive global communication. In *Proceedings of the 13th IEEE International Symposium on Asynchronous Circuits and Systems (ASYNC'07)*, pages 186–195, 2007.

[87] Subhasish Mitra, Wei-Je Huang, Nirmal R.Saxena, Shu-Yi Yu, and Edward J.McCluskey. Reconfigurable architecture for autonomous self-repair. *IEEE Design and Test of Computers, Vol.21, Iss.3*, 2004.

[88] Y. Monnet, M. Renaudin, and R. Leveugle. Asynchronous circuits transient faults sensitivity evaluation. In *Proceedings of the 42nd annual conference on Design automation*, 2005.

[89] Y. Monnet, M. Renaudin, and R. Leveugle. Hardening techniques against transient faults for asynchronous circuits. In *Proceedings of the 11th IEEE International On-Line Testing Symposium (IOLTS'05)*, 2005.

[90] Yannick Monnet, Marc Renaudin, and Regis Leveugle. Designing resistant circuits against malicious faults injection using asynchronous logic. *IEEE Transactions on Computers, Vol. 55, No. 9*, September 2006.

[91] Gordon Moore. Cramming more components onto integrated circuits. Electronics Vol.38, No. 8, 1965.

[92] D.E. Muller and W.S. Bartky. A theory of asynchronous circuits. In *Proceedings of the International Symposium on Theory of Switching*, pages 204–243, 1959.

[93] Suriyaprakash Natarajan, Melvin A. Breuer, and Sandeep K. Gupta. Process variations and their impact on circuit operation. In *Proceedings of the 13th International Symposium on Defect and Fault-Tolerance in VLSI Systems*, 1998.

[94] Eugene Norm. Single event upset at ground level. In *IEEE Transactions on Nuclear Science Vol.43*, pages pp. 2742–2750, 1996.

[95] Patrick D.T. O'Connor. *Practical Reliability Engineering, Fourth Edition*. John Wiley & Sons Ltd., 2002.

[96] Rashad S. Oreifej, Rawad N. Al-Haddad, Heng Tan, and Ronald F. DeMara. Layered approach to intrinsic evolvable hardware using direct bitstream manipulation of virtex ii pro devices. In *Proceedings of the 17th International Conference on Field Programmable Logic and Applications (FPL'07)*, 2007.

[97] Thomas Panhofer, Werner Friesenbichler, and Martin Delvai. Fault tolerant four-state logic by using self-healing cells. In *Proceedings of the 26rd IEEE International Conference on Computer Design (ICCD08)*, 2008.

[98] Thomas Panhofer, Werner Friesenbichler, and Martin Delvai. Optimization concepts for self-healing asynchronous circuits. In *Proceedings of the 12th IEEE Symposium on Design and Diagnostics of Electronic Systems (DDECS09)*, 2009.

[99] N. C. Paver, P. Day, C. Farnsworth, D. L. Jackson, W. A. Lien, and Jianwei Liu. A low-power, low-noise, configurable self-timed dsp. In *Proceedings of the International Symposium on Advanced Research in Asynchronous Circuits and Systems (ASYNC'98)*, pages 32–42, 1998.

[100] Song Peng and Rajit Manohar. Efficient failure detection in pipelined asynchronous circuits. In *Proceedings of the 20th IEEE International Symposium on Defect and Fault-Tolerance in VLSI Systems (DFT 2005)*, 2005.

[101] Song Peng and Rajit Manohar. Fault tolerant asynchronous adder through dynamic self-reconfiguration. In *Proceedings of the 2005 International Conference on Computer Design (ICCD'05)*, 2005.

[102] Song Peng and Rajit Manohar. Self-healing asynchronous arrays. In *Proceedings of the 12th IEEE International Symposium on Asynchronous Circuits and Systems*, ASYNC '06, pages 34–, Washington, DC, USA, 2006. IEEE Computer Society.

[103] O. A. Petlin and S. B. Furber. Built-in self-testing of micropipelines. In *Proceedings of the Third International Symposium on Advanced Research in Asynchronous Circuits and Systems (ASYNC '97)*, April 1997.

[104] P.K.Lala and B.Kiran Kumar. An architecture for self-healing digital systems. In *Proceedings of the Eigth IEEE International On-Line Testing Workshop (IOLTW'02)*, 2002.

[105] P.K.Lala and B.Kiran Kumar. Human immune system inspired architecture for self-healing digital systems. In *Proceedings of the International Symposium on Quality Electronic Design (ISQED'02)*, 2002.

[106] Lucian Prodan, Mihai Udrescu, and Mircea Vladutiu. Self-repairing embryonic memory arrays. In *Proceedings of the 2004 NASA/DoD Conference on Evolution Hardware (EH'04)*, 2004.

[107] Stefan Raaijmakers and Stephan Wong. Run-time partial reconfiguration for removal, placement and routing on the virtex-ii pro. In *Proceedings of the 17th International Conference on Field Programmable Logic and Applications*, 2007.

Bibliography

[108] Marc Renaudin and Yannick Monnet. Asynchronous design: Fault robustness and security characteristics. In *Proceedings of the 12th IEEE International Symposium on On-Line Testing (IOLTS'06)*, 2006.

[109] Gaisler Research. Suitability of reprogrammable FPGAs in space applications. *Feasibility Report FPGA-002-01*, 2002.

[110] R.Kothe, S.Habermann, H.T.Vierhaus, T.Coym, Wolfgang Vermeiren, and Bernd Straube. Selbstreparatur von Logik-Baugruppen in hochintegrierten Schaltungen - M"oglichkeiten und Grenzen. In *Dresdner Arbeitstagung Schaltungs- und Systementwurf (DASS'2006)*, 2006.

[111] Gabi Dreo Rodosek, Kurt Geihs, Hartmut Schmeck, and Stiller Burkhard. Self-healing systems: Foundations and challenges. In Artur Andrzejak, Kurt Geihs, Onn Shehory, and John Wilkes, editors, *Self-Healing and Self-Adaptive Systems*, number 09201 in Dagstuhl Seminar Proceedings, Dagstuhl, Germany, 2009. Schloss Dagstuhl - Leibniz-Zentrum fuer Informatik, Germany.

[112] Goutam Kumar Saha. Software-implemented self-healing system. *CLEI Electronic Journal Vol. 10, No. 2*, December 2007.

[113] Ayodeji O. Olatunji Sharee S. Laster. Autonomic computing: Towards a self-healing system. In *Proceedings of the Spring 2007 American Society for Engineering Education Illinois-Indiana Section Conference*, 2007.

[114] P. Shivakumar, M. Kistler, S.W. Keckler, D. Burger, and L. Alvisi. Modelling the effect of technology trends on the soft error rate of combinational logic. In *Proceedings of the 2002 International Conference on Dependable Systems and Networks*, pages 389–398, 2002.

[115] Martin L. Shooman. *Reliability of Computer Systems and Networks*. John Wiley and Sons, 2001.

[116] A. Holmes Siedle and L. Adams. *Handbook of Radiation Effects*. Oxford University Press, 2002.

[117] S.M.Nowick and D.L. Dill. Synthesis of asynchronous state machines using a local clock. In *Proceedings of the International Conference on Computer Design (ICCD'91)*, pages 192–197, October 1991.

[118] Jens Sparso and Steve Furber. *Principles of Asynchronous Circuit Design - A Systems Perspective*. Kluwer Academic Publishers, 2001.

[119] Mile Stojčev, Teufik Tokič, and Ivan Milentijevič. The limits of semiconductor technology and oncoming challenges in computer microarchitectures and architectures. In *Facta universitatis - series: Electronics and Energetics, vol.17*, 2004.

[120] Ivan E. Sutherland. Micropipelines. *Communications of the ACM*, 32(6):720–738, 1989.

[121] T. Takahara, Y. Kurahashi, T. Mizuno, H. Saito, and N. Tomita. Embedded computer system with soft core cpu for space application. In *Proceedings of the 2002 International Conference on Military and Aerospace Programmable Logic Device (MAPLD'02)*, 2002.

[122] Maxwell Technologies. Scs750 super computer for space. Datasheet 1004741, Rev. 4, 2006.

[123] J. Teifel and R. Manohar. Programmable asynchronous pipeline arrays. In *Proceedings of the 13th International Conference on Field Programmable Logic and Applications*, 2003.

[124] Gianluca Tempesti, Daniel Mange, Pierre-Andre Mudry, Joël Rossier, and Andre Stauffer. Self-replicating hardware for reliability: The embryonics project. *J. Emerg. Technol. Comput. Syst.*, 3, July 2007.

[125] W.B. Toms. *Synthesis of Quasi-Delay-Insensitive Datapath Circuits*. PhD thesis, University of Manchester, 2006.

[126] Davide Tosi. Research perspectives in self-healing systems. *Technical Report LTA:2004:06*, July 2004.

[127] Kees van Berkel, Ronan Burgess, Joep Kessels, Marly Roncken, Frits Schalij, and Ad Peeters. Asynchronous circuits for low power: A dcc error corrector. *IEEE Des. Test*, 11:22–32, April 1994.

[128] VideogeniX. SEUs and Their Effect on Electronic Devices. Whitepaper, 2006.

[129] Sverre Vigander. *Evolutionary Fault Repair of Electronics in Space Applications*. PhD thesis, University of Sussex, Galmer, Brighton, UK, 2001.

[130] Vinu V.Kumar and John Lach. Fine-grained self-healing hardware for large-scale autonomic systems. In *Proceedings of the 14th IEEE International Workshop on Database and Expert Systems Applications (DEXA'03)*, 2003.

[131] Xilinx. Correcting single-event upsets through virtex partial configuration. Application Note 216 v1.0, June 2000.

[132] Xilinx. Two flows for partial reconfiguration: Module based or difference based. Application Note 290 v1.2, September 2004.

[133] Xilinx. Virtex-4 configuration guide. *User Guide UG071 (v1.4)*, January 2006.

[134] Shu-Yi Yu and Edward J.McCluskey. On-line testing and recovery in tmr systems for real-time applications. In *Proceedings of the IEEE International Test Conference 2001 (ITC'01)*, 2001.

i want morebooks!

Buy your books fast and straightforward online - at one of world's fastest growing online book stores! Environmentally sound due to Print-on-Demand technologies.

Buy your books online at
www.get-morebooks.com

Kaufen Sie Ihre Bücher schnell und unkompliziert online – auf einer der am schnellsten wachsenden Buchhandelsplattformen weltweit! Dank Print-On-Demand umwelt- und ressourcenschonend produziert.

Bücher schneller online kaufen
www.morebooks.de

 VDM Verlagsservicegesellschaft mbH
Heinrich-Böcking-Str. 6-8   Telefon: +49 681 3720 174   info@vdm-vsg.de
D - 66121 Saarbrücken   Telefax: +49 681 3720 1749   www.vdm-vsg.de

Printed by Books on Demand GmbH, Norderstedt / Germany